MATH ADVANCEMENT

- ✓Changes in Mathematics Education
- ✓How to Excel on Standardized Exams
- ✓How to Capitalize on Your Unique Learning Style
- ✓An Introduction to Advanced Mathematics
    - Calculus
    - Statistics
    - Probability
    - Chaos Theory
    - Discrete Math

Marilyn K. Simon, Ph.D.

Math Power
Best Prep, LLC
Walden University

KENDALL/HUNT PUBLISHING COMPANY
4050 Westmark Drive Dubuque, Iowa 52002

ISBN 0-7872-5539-4

Printed in the United States of America

10 9 8 7 6 5 4 3 2 1

CONTENTS

Acknowledgments v
Introduction 1
The Ten Commandments of Mathematics 2
The 4R's of Learning 2

PART I: PREPARING FOR STANDARDIZED MATHEMATICAL TESTS

Preparing for Standardized Mathematical Tests 3
Information about Specific Tests 4
Choosing your Own Learning Styles 7
Ancillary Advice 8
To Get in Shape Mentally 9

PRE-TEST 11

Concept 1 : Order of Operation 13
Concept 2: Operations with Fractions 15
Concept 3: Ratios and Proportions 19
Concept 4: Making Comparisons 25
Concept 5: Percents 27
Concept 6: Square Roots 29
Concept 7: Solving Equations and Inequalities 31
Concept 8: The Distributive Property 34
Concept 9: Averages 39

Check Point # 1 43

Concept 10: Equal Angles 47
Concept 11: Area and Perimeter 51
Concept 12: The Circle 55
Concept 13: Angles and Circles 57
Concept 14: The Pythagorean Theorem 59

Check Point # 2 63

The Other 10 Percent 65
Indirect Methods of Solving + Guessing Strategies 69
More Problems to Ponder 71
Data Sufficiency (GMAT™) 75
Quantitative Comparisons (SAT™ and GRE® Test) 81

Data Interpretation 83
Who Owns The Zebra? 85
Analytical Reasoning and Logical Puzzles (GRE® Test) 87
If P then Q: $P \rightarrow Q$ 91
More Tips on Logical Puzzles 99

Post Test 107

PART II: MATHEMATICS TODAY AND ADVANCED MATHEMATICS

What is the Probability of? 113
What's Stat you Say? 121
Calculating Calculus 123
Chaos Theory and Fractal Geometry 125
Delightfully Discrete Mathematics 129
Math Powers 133
What's Your Preference 135
Words and Ideas for Math Stars 137
Exercises for Test Takers 139
Helpful Hints 141
Feedback to Best Prep Instructors 145

Answer Key 147

ACKNOWLEDGMENTS

I wish to acknowledge and thank the following people:

- Mark Peterson for his creative and humorous designs;
- Professor Dorothy Geddes for her invaluable mathematics education program;
- my BEST-PREP partner, Scott J. Bornstein; and
- all our wonderful BEST-PREP students.

In particular, I would like to express my deepest appreciation to my husband, Dr. Ronald Alan Simon, and our sons, Matthew and Jonathon, for their support and encouragement to me on this project.

Finally, for all those who have come to see the beauty of mathematics, AH MATH!!!

INTRODUCTION

Our world is becoming more mathematical. We are constantly surrounded by circumstances which require mathematical thinking and decision making. These decisions require number sense, logic, the ability to analyze data, knowledge of geometry, and many other abilities and skills not often taught in primary or secondary schools.

Anyone who has read the media reports about mathematics education, heard about the several comparative international studies, or is aware of the analyses of the National Assessment of Educational Progress (1998) is likely to conclude that we are not doing a good enough job to help our populace learn and understand enough mathematics to lead productive and fulfilling lives in a modern society. The kind of mathematics that people need in the age of computers and calculators is different from the mathematics needed in the past. There is a need to de-emphasize the teaching of computational skills and to stress the development of mathematical thinking and problem-solving. The world is becoming more complex and more quantitative. As proclaimed by The National Council of Teachers of Mathematics (NCTM) in the *Curriculum and Evaluation Standards* (NCTM Standards, 1989) there is a need for a mathematically literate society, a flexible workforce, mathematics equity in education, and a technologically informed electorate.

Entrance into higher education, or obtaining professional licensure, often requires examinations in which mathematical knowledge is tested. Familiarization with test structure, content of the questions, and test strategy will lessen anxiety about these tests, and ultimately help with performance. Preparing for a test builds the confidence necessary to perform successfully.

Math Advancement is designed to give you the information you need to understand the mathematics sections of important exams, and to introduce you to the basic concepts of advanced mathematical subjects that are usually not a part of the primary and secondary curricula, but are often required in higher education and in solving real-world problems. Such subjects include discrete mathematics, probability and statistics, chaos theory, and calculus.

In the *Magical Maze* (1998) Ian Stewart explains that mathematical thinking "involves navigating, with confidence, through an intricate network of logical possibilities [and that] mathematics is the closest that we humans get to true magic (p. 2)." It is the intent of *Math Advancement* to guide you through the mathematical maze of concepts you will encounter on standardized exams and to introduce you to the beauty and magic of mathematical thinking.

The Ten Commandments of Mathematics from the MATHIAH

1. Thou shalt never divide by zero.
2. Thou shalt do unto one side of an equation as thou does unto the other.
3. Thou shalt think oneness when thy numerator (up) and denominator (down) are equivalent.
4. Thou shalt always think "out of 100" or "over 100" when thou sees "%."
5. Thou shalt obey the order of operations by remembering the Girl Scout who gave power to my Dear Aunt Sally.
6. Thou shalt honor thy variables' request to be alone by undoing what has been done.
7. Thou shalt distribute fairly.
8. Thou shalt use thy intuition.
9. Thou shalt help thy neighbor DO MATH, except during most tests.
10. Thou shalt never say: "I can't do math," or "I hate math." Positive thinking only!

Ah, Math!!

Remember the Four R's of Learning

1. **Responsibility**: Acknowledge your role in learning. That gives you real power to effect change and grow.
2. **Re-view your script**. If you got lost on a trip you would: re-check your map, find out where you went wrong, and get back on the trail. When you make a mistake, go back in your mind and re-view the situation. Did you make a careless error? Miss-read the question? Miss-use a concept? Look at both your behavior and the thought that lead to the error.
3. **Re-place the old footage**. To ensure you don't make the same mistake again, replace the old footage with the new improved one.
4. **Re-hearse the new plan**. The desired behavior when well rehearsed will become automatic and will convince you that you CAN DO MATH.

PREPARING FOR STANDARDIZED MATHEMATICAL TESTS

Congratulations! You have decided to prepare for your test, and because of this you will have an edge in achieving the best possible score. When preparing for any exam, it is an excellent idea to discover as much as you can about its content and context. The mathematics section of most standardized tests includes topics ranging from arithmetic to algebra and geometry. You will be tested on your ability to apply certain concepts in a variety of situations. It is likely that when you studied these concepts in school they were not presented in the same way that you will see them presented on your test.

How to Prepare for Your Test Using Math Advancement

There are 14 basic mathematical concepts presented in **Part I** of ***Math Advancement***. You will see that each concept has a full-page illustration associated with that concept. By studying each illustration and making a mental association, you will reinforce a method for successfully answering the questions on your test. This will help to transfer the information into the long-term computer of your mind. At least 90 percent of test questions on standardized exams in mathematics include these 14 concepts. Take time to review the concepts and you will be able to quickly answer these questions with ease and confidence.

Math Advancement also covers special mathematical sections on the exams such as analytical reasoning, quantitative comparisons, data analysis, data sufficiency, and logical puzzles.

In addition, you will find alternative problem-solving methods and effective strategies for guessing.

As you review for your test, use the checkpoints in ***Math Advancement*** to keep track of your progress. Do not be concerned about making a mistake during your preparation. Learn from your mistakes. It is much better to make a mistake during your preparation than on the test itself.

If you have another way to solve a problem quickly that is different from the method presented in ***Math Advancement***, then *more power to you!* One of the wonderful properties of mathematics is that there are so many ways to arrive at a solution to a problem. The more ways you learn, the better problem-solver you become. Learning alternative ways also gives you a means of checking so that you are 100 percent sure of your answer.

What Inquiring Minds Will Want to Ask Themselves

When encountering a new concept in **Part I** of ***Math Advancement***:

> *"What do I already know about this concept?"*
> *"When have I used this concept before?"*

When reading a problem:

> *"What do I need to find?"*
> *"What information is given?"*
> *"How can this be put another way?"*

After solving a problem:

"What did I do right?"
"What misunderstandings did I have?"
"How could this problem be altered to create another problem?"
"How could I explain this problem to someone else?"

On most, but not all, standardized exams in mathematics each question is worth the same. A few questions are extremely easy, while some are extremely challenging. Most questions fall in the middle of the two extremes and become easy once you know which concepts to apply.

You have about one minute per question. You are graded on the number of questions you get correct, so it is the author's recommendation that you do not leave any blanks. The guessing strategies in ***Math Advancement*** work very well for challenging problems and for computer adaptive tests.

Although memory is an important aspect of learning, mathematics is more than a set of rules to be memorized. ***Math Advancement*** offers you an opportunity to, among other things, investigate patterns, understand concepts, formulate hunches, and construct your own understanding of mathematics. In the *Math Powers* section you will find open-ended problems that will help further your mastery of mathematics.

It is recommended that you relax before and during the test. Try to reduce the level of physical and mental stress associated with the test. The *Helpful Hints and Exercises for Test Takers* will help prepare you for a successful performance.

Information About Specific Tests

The SAT-I™ is a three-hour multiple-choice test that is divided into seven sections:

- two 30-minute verbal sections
- one 15-minute verbal section
- two 30-minute math sections
- one 15-minute math section
- one 30-minute experimental section, either math or verbal

The 15-minute verbal section consists of one double critical reading passage followed by 13 questions. The 15-minute math section consists of 10 multiple-choice math questions. Only the verbal and math sections will count toward your SAT™ scores. The experimental section on your SAT™ may look just like a verbal section or a math section, but it won't be scored; ETS™ uses it only to try out new SAT™ questions and to determine whether the test you are taking is harder or easier than ones that have been given in the past.

The verbal SAT-1™ contains three types of questions.

1. Sentence completions.
2. Analogies.
3. Critical reading.

The math SAT-1™ also contains three types of questions.

1. Regular multiple-choice math (arithmetic, algebra, and geometry).
2. Quantitative comparisons.
3. Grid-ins.

The Preliminary Scholastic Assessment Test (PSAT™) is usually administered in eleventh grade, although ninth- and tenth-graders also take this test, as do some eighth graders. The PSAT™ is arranged a little differently from the SAT-1™. It contains two 30-minute verbal sections and two 30-minute math sections. (There is no experimental section on the PSAT™.) Questions on the PSAT™ are taken directly from old SAT-1™ exams.

On March 27, 1992, The College Board approved the implementation of calculators on the mathematics section on the PSAT/NMSQT™ that was administered in October 1993, and on the new SAT-1™ in March of 1994. These test takers are permitted to use **any** four function, scientific or graphing calculator. However, no question requires the use of a calculator.

The mathematics sections also requires students to produce their own response to some questions, rather than select all answers from a set of multiple choice alternatives. These "**grid-in**" items make up about 20% of the mathematics test.

The content and the concepts are in accord with the content and concepts you find in ***Math Advancement***. Students have 70-75 minutes to answer 60 questions. There are 35 multiple choice questions, 15 quantitative comparisons and 10 "grid-ins." Throughout ***Math Advancement*** you will find multiple choice and quantitative comparison questions as well as questions asking you to produce your own answers. What follows are some sample "grid-in" questions and the ***Math Advancement*** concept in Part I that they correspond to.

1. If Henrietta has an average of 80 on four tests, what can she get on her fifth test to raise her average to be greater than 81 and less than 82. Test scores are integer values. [concept 9]

2. In triangle ABC, point D (not shown) is between B and C. Grid in the possible values for the length of AD. [concept 10]

3. Grid in a value for x if $\frac{1}{2} < x < \frac{2}{3}$ [concepts 2,3]

4. In triangle ABC, the measure of angle A is 80 degrees and angle B is the smallest angle. If angle B is larger than 1/2 the other angles of the triangle, what is a possible integer value of angle C. [concept 10]

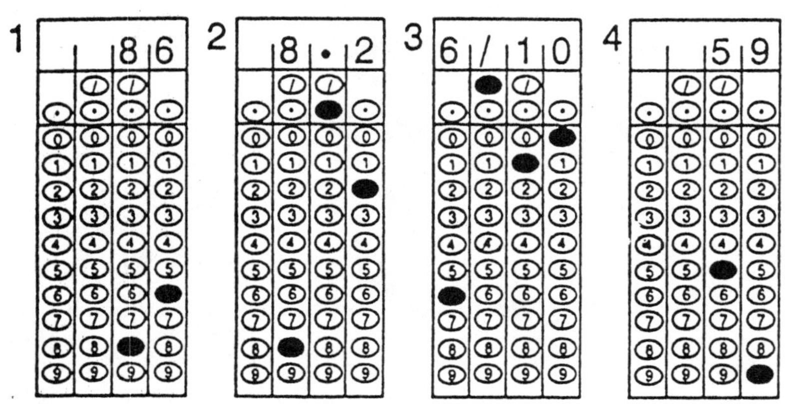

(also 87, 88, or 89) (any value between 8 and 9) (any value between 0.5 and 0.666)

9

Where Do Most of These Tests Come From?

The P/SAT™, GRE® Test, GMAT™, CBEST™, and Praxis portion of the MSAT™ are published by the Educational Testing Service (ETS™) in Princeton, New Jersey, under the sponsorship of the College Entrance Examination Board (College Board). ETS™ and the College Board are both private non-profit companies.

How Important Are These Scores?

Entrance exams are an important factor when you apply to colleges, but they are not the only one. In general, the larger the college, the more important the score. Small liberal arts colleges and most places with highly selective admissions processes will give a great deal of weight to your extracurricular activities, interviews, essays, and recommendations. Many public and large private universities admit students based on formulas consisting of just two ingredients: entrance exam scores and grade point average. If your scores are good, many admissions officers will be more likely to give you the benefit of the doubt in other areas. Many nontraditional schools do not put much weight on these exams.

Computer Adaptive Tests

The GRE® Test and GMAT™ are for students seeking admission into graduate schools. Both tests now provide Computer Adaptive Tests (CAT™). The CAT™ differs from the paper-pencil based tests in that a computer program chooses problems based on your response to a previous question. Your score is dependent on how well you do on the question presented as well as the number of questions you answer.

In a CAT™, you must answer every question in the order in which it is given. Usually the first question is of "middle" difficulty. If the easiest question was ranked "0" and the most challenging was ranked "10," then the first question would be ranked "5." Your answer to previous questions helps to determine the next question that will be asked. Thus, in a CAT™:

- One question appears at a time. You CANNOT go back.
- To answer you click on an OVAL rather than a letter.
- Once you have confirmed your answer you automatically move to the next question.
- There are generally fewer questions on the CAT™ than on a paper and pencil test.
- Questions are weighted. The more challenging questions have a higher weight.
- You can choose to see your score instantly upon completion of the test or cancel them at that time prior to seeing the score.
- You can take the CAT™ year-round at your convenience.
- There is usually a 10-minute warm-up period to get comfortable with the hardware and the software.

Choosing Your Own Learning Style

Math Advancement will help you learn how to formalize a bridge from your instinctive understanding of mathematics to formal communications in mathematics. There are four major learning styles appearing in **Part I** of ***Math Advancement*** that can be referred to as the 4**P**'s: **P**lunging, **P**rocessing, **P**icturing, and **P**racticing. You will be able to learn the information in ***Math Advancement*** by using any combination of the 4**P**'s.

"Plungers" prefer to delve right into a problem and devise their own techniques for solving it. They are usually risk takers and are often frustrated by what they view as excess information because they trust their instincts and wish to apply them as quickly as possible. If you are a ***Math Advancement*** student and see yourself as a *plunger*, you might wish to do the problems relating to a concept first and then try reading the information about the concept if you find any obstacles in obtaining a solution.

"Processors" prefer a recipe or a systematic means by which they will be able to successfully get to the answer. They are not fond of false starts or loose structure and instead prefer to have a set plan before *plunging* into solving a problem. When a new concept is introduced in ***Math Advancement***, appearing on the right side of the text is a step-by-step procedure to solve problems relating to the particular concept. If you are a processor you might wish to read the given information before starting on the problem sets.

"Picturers" prefer to take time to perceive information as a whole before attempting to apply that information. They often seek to categorize information in a way that correlates with other information that they have. If you see yourself as a *picturer*, then you might wish to study the visual representation of the concept first. You will find these on the left side of the text. After you feel you have a general understanding, you might wish to read the information about the concept or try solving the related problems.

"Practicers" prefer to see a model problem presented first before attempting independent work. They prefer a validation that the information is applicable to a problem and will lead to a correct solution, should they decide to invest their time in solving it. If you are a *practicer* you might wish to check the sample problems that have been "squared off" before attempting to solve any problems on your own.

Allowing for diversity in learning is an essential ingredient of ***Math Advancement*** and has been included so that you will have an easier time relating to the material. Once you are aware of your own preference, and the preferences of others, you might wish to try learning a concept by using a learning style that is not necessarily your preferred one. Learning math in a variety of ways helps to expedite your mastery and advance in mathematics.

Ancillary Advice

- Answer the question that is asked - not the one that you *think* they are asking. You might wish to underline words that come after what is (are), how many, which is (are).
- The easiest questions are usually the first ten. Don't reject the "obvious" here.
- The last ten questions are usually the most challenging. *Always* question the "obvious" answer here.
- If your test has *grid-in* answers, then the largest *grid-in answer* is 9999 and the largest decimal *grid-in* is 99.9. There are no negative *grid-in* answers, and you cannot *grid-in* π If your answer has a % or $ you will be told to omit these symbols.
- To check quantitative comparison questions involving variables you might wish to check: 0,–1,0.5, 1, and 2 before making a generalization.
- Mathematics is very literal. Check the meaning of each word in the context of the problem. For example, the word "result" could mean different things in different questions.
- Even if you *are* allowed to use a calculator on your test, many times reducing answers can lead to a quicker solution than working the problem through with the aid of a calculator. If you are allowed to use a calculator, make sure you use a calculator that you are comfortable with. Calculators that convert decimals to fractions might be helpful. It is also a good idea to purchase a calculator that has ( ) and a two line read out. These are usually scientific calculators and sell for under $15.

- If the sum of the digits are divisible by 3, then so is the entire number. The same is true for 9. Thus: 634,275 has both 3 and 9 as factors.

- Always think positively!!!

To Get In Shape Mentally

It seems that some people's minds are sharp, energetic, and flowing with creative ideas while other people's intellect seem less than dazzling. The reason for this discrepancy appears to be by circumstances and mental habits.

A construction worker does not need to go to the gym to build up his/her muscle strength. In the course of his/her work (circumstance) his/her arms are strengthened. A financial planner does not need to review arithmetic skills because he/she is dealing with numbers all the time.

If your circumstances don't challenge your thinking muscles, you might not be getting the work out you need. The following activities can help you sharpen the muscle between your ears, and thus enable you to perform to your optimum on a test in logical reasoning:

- Crossword puzzles
- Visualizing the face of a teacher you had in grade school
- Finding a hundred uses for a paper clip
- Drawing a Picasso that is turned upside down
- Doing a 500 piece jigsaw puzzle
- Learning a new computer program
- Listing five concerns you would have if you won the lottery
- Recalling a conversation you had last month
- Planning a trip to some place you have never been before
- Learning the words to a new song
- Reciting the alphabet backwards

Just as athletes watch their diet to enhance their physical output, your diet can also help you to increase your mental output. Protein helps the brain stay alert and make new synapses. Choline is also believed to be a key factor in memory. Choline is found in egg yolks, beef liver, fish, legumes and raisins. Dietary iron plays a vital role in getting oxygen to the brain. Beef and soybeans are good sources of iron. Vitamin C (found in most fruits and vegetables) increase the bodies ability to absorb iron.

In addition to diet, physical exercise on a regular basis (at least 20 minutes of aerobic exercise three days a week) have profound benefits for memory improvement, general intelligence, creativity, and emotions.
Sleep is also important for the brain. Most Americans do not get enough sleep. A poor night's sleep can make itself felt when doing tasks that require mental clarity. Most adults need about 8 hours of sleep to function optimally. Along with sleep, meditation can have major benefits for mental well being. Meditation includes many types of activities from simply collecting one's thoughts, to positive visualization, to a highly structured form of Eastern meditation and relaxation.

Remember the BEST thing you can bring to a test is self confidence. You deserve to pass! You are a Best-Prep Student!

WARM-UP ACTIVITY

See if you can interpret these Plexirs [answers on bottom].

1. $\frac{\text{Long}}{\text{Due}}$
2. Every right thing
3. death life
4. yyymen
5. DKI
6. tom 000 000 00
7. dice dice
8. $\frac{\text{mind}}{\text{matter}}$
9. $\frac{\text{iiiiii}}{\text{000000}}$
10. Mill1ion
11. $\frac{\text{egg}}{\text{EZ}}$
12. $\frac{\text{Man}}{\text{board}}$
13. cycle cycle
14. | R | E | A | D | I | N | G |
15. ECNALG
16. LE_{VEL}
17. $\frac{\text{WEAR}}{\text{LONG}}$
18. $\frac{\text{GI}}{\text{CCC}}$
19. T
O
W
N
20. T
O
U
C
H
21. $\frac{0}{\text{M.D. PH.D. M.S.}}$
22. $\frac{(90 - C_1), (90 - C_2), \ldots (90 - C_n)}{2U}$

Answers: 1. long over due 2. right in the middle of everything 3. life after death 4. three wise men 5. mixed up kids 6. tomatoes 7. paradise 8. mind over matter 9. circles under the eyes 10. One in a million 11. Eggs over easy 12. Man overboard 13. bicycle 14. reading between the lines 15. sideways glance 16. split level 17. long underwear 18. GI over seas 19. Downtown 20. Touchdown 21. 3 degrees below zero 22. The compliment of the seasons to you.

MATH MASTERY

1. In the figure, PQRS is a rectangle, and ST = 1/4 RS. If ΔPST has an area of 15, what is the area of the rectangle?

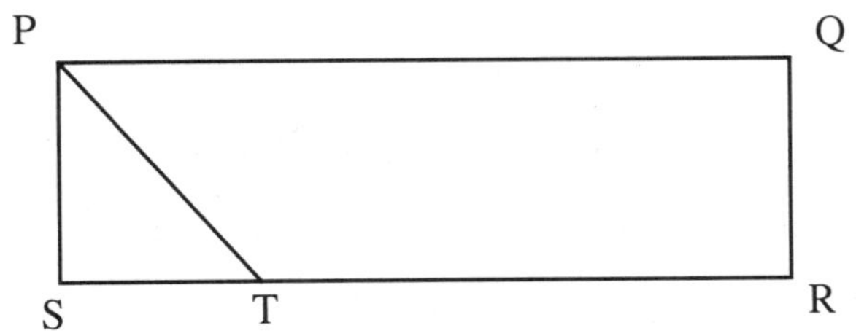

a) 60 b) 75 c) 90 d) 120 e) 140

2. A carpet is y yards long and f feet wide. How many dollars will it cost, if the carpet sells for x cents per square foot?

a) xyf b) 3xyf c) $\frac{xyf}{3}$ d) $\frac{.03yf}{x}$ e) .03yfx

3. A square is inscribed in a circle of area 18π. A side of the square is:

a) 3 b) 6 c) $3\sqrt{2}$ d) $6\sqrt{2}$ e) cannot be determined

4. A man buys a radio for $72 after receiving a discount of 20%. What is the ratio of the sale price to the list price?

a) 1 : 5 b) 2 : 5 c) 3 : 5 d) 4 : 5 e) 3 : 4

5. If one side of a square is increased by 3 and another side decreased by 3, a rectangle is formed whose area is 40. What is the area of the original square?

a) 31 b) 34 c) 37 d) 43 e) 49

6. How many square units in the area of the semicircle in the figure below?

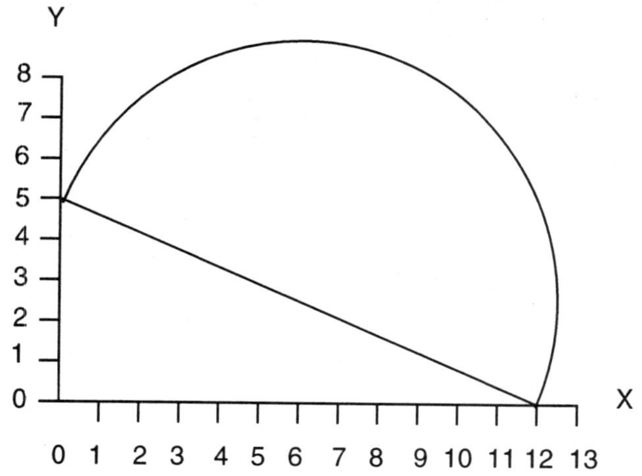

a) 169π b) $\frac{169\pi}{4}$ c) $\frac{169\pi}{8}$ d) $\frac{169\pi}{16}$ e) $\frac{13\pi}{2}$

7. In a circle with center at O, arc RS = 132°. How many degrees in angle RSO?

a) 18 b) 20 c) 22 d) 24 3) 26

8. In the formula $y = x^2 + rx - 3$, for what value of r will y = 11 when x = 2?

a) 6 b) 5 c) 4 d) $3\frac{1}{2}$ e) 0

9. If t tons of snow fall in h hours, how many tons will fall at this rate in m minutes?

a) $\frac{mt}{h}$ b) $\frac{mh}{t}$ c) $\frac{60mt}{h}$ d) $\frac{60t}{mh}$ e) $\frac{mt}{60h}$

10. A gear 50 inches in diameter turns a smaller gear 30 inches in diameter. If the larger gear makes 15 revolutions, how many revolutions does the smaller gear make in that time?

a) 9 b) 12 c) 20 d) 25

CORRECT

$\frac{\square}{10}$

ORDER OF OPERATIONS

Many problems on your exam will require performing more than one mathematical operation. Depending on the order you choose to evaluate these operations, your answer will vary. Since there is only one correct answer, it is crucial that you follow the *"Girl Scouts Simple Order of Operations."* When operations have the same rank, "yield to the left," that is, do what's on the left first. The Girl Scouts Simple Order is:

1. The **GIRL SCOUT** — Grouping symbols are evaluated first. Grouping symbols come in pairs. They may be parentheses ( ), brackets [ ], or braces { }. If there is more than one set evaluate the innermost first. Think of a Magician who is inside a paper bag and put into a suitcase. He needs to get out of the bag first and then he can get out of the suitcase.

2. Who **GAVE POWER** — Powers, or exponents, are ranked second.

3. To **MY DEAR** — Multiplication and division are both ranked third.

4. **AUNT SALLY** — Addition and subtraction are both ranked last. We save the easiest operations for last.

And got a good EVALUATION.

EVALUATE $5 + 2^3 \div 4 + (1 + 2)^2 \times 2 - 5^2$

ANSWER

1. $5 + 2^3 \div 4 + 3^2 \times 2 - 5^2 =$
2. $5 + 8 \div 4 + 9 \times 2 - 25 =$
3. $5 + 2 + 18 - 25 = 0$

TRY THESE

1. If $K = -1$ then $K + K^2 + K^3 + K^4 =$

   a) -4 b) -2 c) 0 d) 2 e) 4

2. If for all real numbers a and b, $a \# b = a^2 - 2ab$ then $3 \# 2 =$

   a) -5 b) -13 c) -3 d) 21 e) 3

3. If for all real numbers $(a, b, c * x, y, z) = ax + by + cz$, then $(2, 2, 5 * 1, -1, 3) =$

   a) 0 b) 30 c) 15 d) -15 e) 5

OPERATIONS WITH FRACTIONS

It is likely that you will be called upon to add, subtract, multiply and divide fractions on your exam. You have probably added and subtracted fractions by finding common denominators in the past:

$$\frac{2}{9}+\frac{1}{3} = \frac{2}{9}+\frac{3}{9} = \frac{5}{9}$$

If a common denominator cannot be found at a glance, then a quick and easy way to add (subtract) fractions is to:

$$\frac{\text{NOD} \pm \text{NOD}}{\text{DEN} \cdot \text{DEN}}$$

1. Multiply the first **N**umerator by the **O**pposite **D**enominator (**NOD**).
2. Add to (subtract from) this the product of the second **N**umerator times its **O**pposite **D**enominator (± **NOD**).
3. Make this sum (difference) the numerator of your answer.
4. Make the **DEN**ominator of the answer the product of the two original **DEN**ominators (**DEN · DEN**).

OBSERVE $\frac{a}{b} + \frac{c}{d} = \frac{ad + bc}{bd}$ when b and d $\neq 0$

and

$$\frac{2}{3} + \frac{4}{5} = \frac{10 + 12}{15} = \frac{22}{15} \text{ or}$$

$$\frac{2}{3} - \frac{4}{5} = \frac{10 - 12}{15} = \frac{-2}{15}$$

Notice, you are performing 3 multiplications and one (addition/subtraction) when you (add/subtract) two fractions.

The fraction bar serves as a division symbol and as a grouping symbol. When variables are in denominators make sure that the denominator could not possibly be zero. The first commandment of mathematics states:

Thou shalt never divide by zero.

The easiest operation to perform with fractions is multiplication.

OBSERVE If b and d $\neq 0$ then, $\frac{a}{b} \cdot \frac{c}{d} = \frac{ac}{bd}$

Multiplying Fractions is never a problem, when you multiply the tops and multiply the bottoms.

That is, when multiplying fractions simply — **"Squish"** — multiply the numerators and — **"Squash"** — multiply the denominators. (In multiplication, you are allowed to reduce diagonally.)

$$\frac{\overset{1}{\cancel{3}}}{\underset{2}{\cancel{22}}} \times \frac{\overset{1}{\cancel{11}}}{\underset{9}{\cancel{27}}} = \frac{1}{18}$$

Dividing fractions is almost as easy as multiplying. Instead of going Squish/Squash, go Squash/Squish and make a little fish.

$$\frac{3}{4} \div \frac{2}{7} = \frac{3 \times 7}{4 \times 2} = \frac{21}{8}$$

Notice the second fraction is "flipped" over when dividing. (In division, you are allowed to reduce across.)

$$\frac{\overset{1}{\cancel{3}}}{10} \div \frac{\overset{2}{\cancel{6}}}{7} = \frac{7}{20}$$

SIMPLIFY $\left(\frac{3}{4} \times \frac{1}{5}\right) \div \left(\frac{4}{5} - \frac{3}{4}\right)$

ANSWER $\frac{3}{4} \times \frac{1}{5} = \frac{3}{20}$ (Squish/Squash)

$\frac{4}{5} - \frac{3}{4} = \frac{16 - 15}{20} = \frac{1}{20}$ $\frac{(\text{NOD} + \text{NOD})}{(\text{DEN} \cdot \text{DEN})}$

$\frac{3}{20} \div \frac{1}{20} = \frac{60}{20} = 3$ (Squash/Squish)

TRY THESE

1. Simplify $\frac{1}{m} - \frac{1}{n}$, m and $n \neq 0$

2. $\frac{7}{15} \times \frac{9}{14} \div \frac{9}{10} =$

   a) $\frac{1}{6}$ b) $\frac{1}{3}$ c) $\frac{1}{2}$ d) $\frac{3}{5}$ e) $\frac{2}{3}$

3. Initially there are 18 bananas on a tree. One monkey eats 1/3 of the bananas and another eats 1/3 of what is left. How many bananas are still hanging in there?

   a) 4 b) 6 c) 8 d) 10 e) none of these

4. $\dfrac{\frac{3}{10} - \frac{1}{2}}{\frac{1}{4} - \frac{1}{5}} =$

5. The operation # is defined by the equation: $x \# y = \frac{x - y}{2x - y}$ where $2x \neq y$. Which of the following is equal to 0?

   a) 2 # 2 b) 2 # –2 c) 2 # 2 d) 1 # 2 e) 3 # 2

6. If $\frac{X}{Y} > 1$ which of the following must be less than 1:

a) $\frac{2Y}{X}$ b) $\frac{X}{2Y}$ c) $\frac{X}{Y}$ d) $\frac{Y}{X}$ e) $\frac{X^2}{Y^2}$

7. If a pipe fills a pool in 5 hours and another empties it in 6 hours, if both pipes are left on by mistake for 3 hours when the pool was empty, how much of the pool will be filled?

a) 1/10 b) 3/5 c) 3/10 d) 3/11 e) 1/6

8. If $m = 1/2$ then $m^2 - m + 1/4 =$

a) 0 b) 1/4 c) 1/2 d) 3/4 e) 1

CONCEPT 3

RATIOS AND PROPORTIONS

The concept of ratios and proportions appears in all areas of mathematics. Perhaps this is why it appears so often on standardized exams. Ratios give us a way to compare two quantities. A ratio may be thought of as a:

	Fraction	Comparison	Rate	Percent	Decimal	Quotient
THINK:	**FAST**	**CATS**	**RUN**	**PRETTY**	**DARN**	**QUICK**

Percents compare a quantity to 100 of that quantity; decimals compare quantities to powers of 10, i.e. 10^2, 10^3, . . ., 10^n. [if n is a whole number then the power equals the zeroes following the 1, i.e. $10^{\mathbf{4}}$ = **10,000;** $10^{\mathbf{0}}$ = 1]; Rates compare quantities to **1**.

To change from a **fraction** to a **decimal** just divide numerator by denominator.
To change a **decimal** to a **percent**, just multiply by 100 (move decimal point two places to the right) and attach %: **1/4 = 0.25 = 25%.** If you are asked to compare two quantities and express your answer as a ratio,

make certain you give them the ratio they are asking for. If Jon missed 3 questions and got 7 correct, then the part of the questions he missed was 3/10, or 0.3 or 30%.

Other important ratios to know include:

Slope: a ratio that tells you the *steepness* and the direction of a line. Slope is the ratio of the rise to the run between any two points on the line. The larger the slope the greater the rate of change. Positive slopes point Northeast.

Probability: is a ratio between 0 (impossible) and 1 (certain).

If in a given experiment (pick a card) there are N equally likely outcomes (52 different cards, from a standard deck), and j ways that an event J can occur (there are 4 ways to get a jack) , then the probability of J = j /N (4/52 or 1/13).

A proportion is an equation between two ratios:

$$\frac{2}{3} = \frac{4}{6}$$

If you find 4 different numbers that form a proportion, you can actually form 4 true proportions with these numbers. However, there are 12 ways to arrange these numbers that will not give true proportions. To solve problems we will use a method that yields one of these true proportions. (You might have used another method of solving that involves a different proportion.)

EVALUATE Using the numbers (2, 3, 4, and 6), make 4 proportions.

ANSWER Since $2 \times 6 = 3 \times 4$, we must make sure that 2 and 6 are diagonal to each other. Also, 3 and 4 will be diagonal.

a) $\frac{2}{3} = \frac{4}{6}$ b) $\frac{3}{2} = \frac{6}{4}$ c) $\frac{2}{4} = \frac{3}{6}$ d) $\frac{4}{2} = \frac{6}{3}$

TRY THIS

☞ Make 4 proportions with the set of numbers (2, 10, 5, 4).

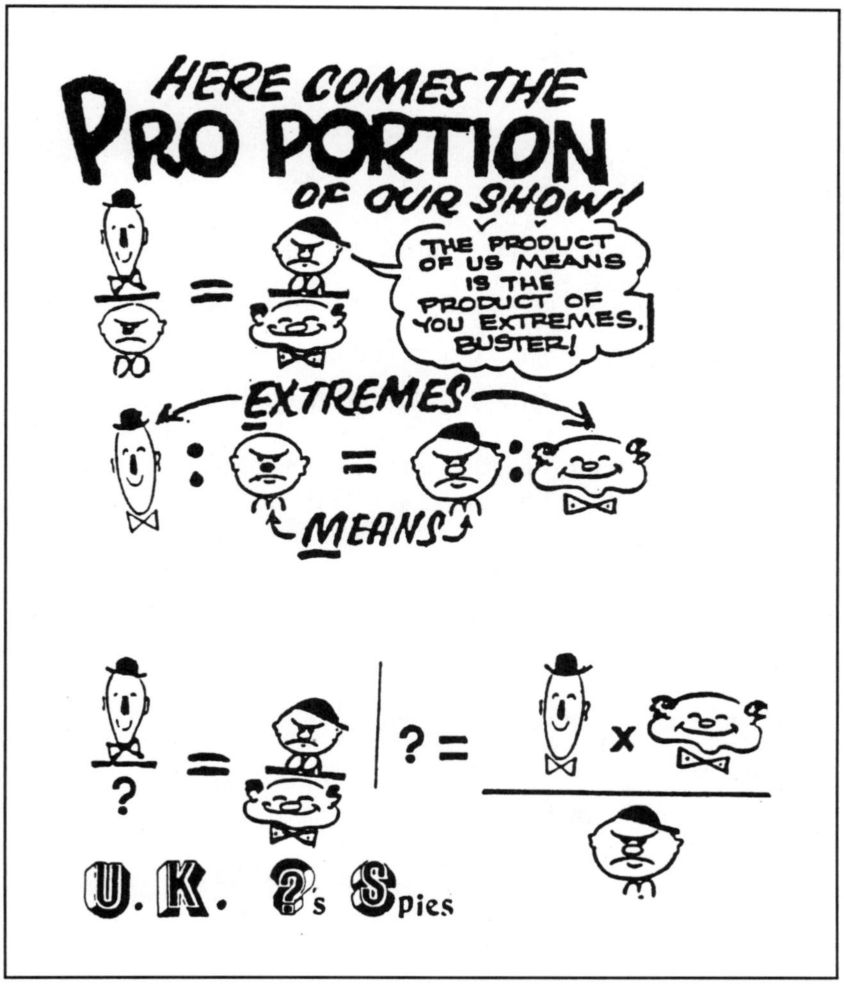

Solving Proportions

We refer to the terms of a proportion as the EXTREMES and the MEANS. In the proportion:

$$\frac{a}{b} = \frac{c}{d} \text{ or } a : b = c : d$$

(MEANS: b and c; EXTREMES: a and d)

a and d are the **EXTREMES** on the end and b and c are the **MEANS** in the middle. The fundamental theorem of arithmetic (a very important statement that can be proven true) states:

In a proportion, the product of the extremes equals the product of the means.

OBSERVE $\frac{2}{3} = \frac{4}{6}$ and $\frac{2 \times 6}{12} = \frac{4 \times 3}{12}$

$$\text{If } \frac{7}{10} = \frac{3}{x} \text{ then } 7x = 30 \text{ and } x = \frac{30}{7} \text{ or } 4\frac{2}{7}$$

To solve a proportion means to find one missing term when 3 terms of a proportion are known. You can do this quickly by

1. Multiplying the pair (means or extremes) that's given and then,
2. Dividing this product by the other given term (i.e., the term diagonal to the missing term).

If a proportion involves variables,
it is an excellent idea to use a "?" for the missing term.

It is also an excellent idea to reduce a proportion before multiplying and dividing. A proportion can be reduced if you can find a number that divides (is a factor of) any pair of numbers straight across (horizontally) or up and down (vertically).

$\frac{24}{9} = \frac{80}{x}$ can be reduced by dividing 24 and 80 by 8, obtaining

$\frac{3}{9} = \frac{10}{x}$ which can then be reduced further by dividing 3 and 9 by 3,

$\frac{1}{3} = \frac{10}{x}$ which easily gives us the answer x = 30.

TRY THESE

Find the missing term (assume all variables are not zero).

1. $\frac{2}{7} = \frac{x}{11}$
2. $\frac{7}{9} = \frac{4}{y}$
3. $\frac{a}{?} = \frac{b}{m}$
4. $\frac{Rx}{10T} = \frac{?}{Q}$
5. $\frac{33}{11} = \frac{69}{w}$
6. $\frac{32}{80} = \frac{x}{5}$

Solving Word Problems

Many problems on your exam may be solved by setting up a proportion. When the problem involves units (i.e., quantities being measured or compared) such as feet, miles, pounds, people, hours, lightbulbs, cars, etc., there is an excellent chance that we can apply an easy method to solve the problem. First, decide if the quantities are directly or inversely related and think:

THE UNITED KINGDOM QUESTION SPIES

Directly proportional — Quantities are growing or shrinking at the same rate.

EXAMPLE If M tons of snow falls in V hours, how much snow will fall in T hours?

1. **UNITS (UNITED)**
   Identify 2 units from the problem and write them at the top of the two columns.

TONS	HOURS

2. **KNOWNS (KINGDOM)**
   Write a known relationship in the appropriate columns. (This might be a fact that you need to supply, such as 1 foot = 12 inches or will be given in the problem.)

TONS	HOURS
M	= V

3. **UNKNOWN (QUESTIONS?)**
   Identify what is missing with a question mark (?) and what it is equivalent to. Put this information in the appropriate columns.

TONS	HOURS
$\frac{M}{?}$	= $\frac{V}{T}$

4. **SOLVE (SPIES)**
   Solve the proportion.

$$? = \frac{MT}{V}$$

Inversely proportional - As one quantity increases, the other decreases (like a seesaw). Here steps 2 and 3 will be slightly altered. The known quantities will appear diagonal to each other (like a seesaw) instead of across from each other.

EXAMPLE If it takes 5 men 3 days to do a job, how long would it take 10 men?

↑ MEN	DAYS ↓
$\frac{5}{10}$	$= \frac{?}{3}$

? = 1.5 days - (more men, less days)

Note: The UK method can be used whenever two quantities, measured in various units, are being compared in such a way that the increase of one of these quantities implies a proportional increase or decrease in the other quantity. Only 2 different units are permitted in this method. Once the proportion is established, look for the easiest way to solve it!

TRY THESE

1. If 8 oz of milk contains 180 calories, how many calories will there be in 12 oz?

   a) 120 b) 240 c) 360 d) 270 e) 300

2. If Martin types at 45 wpm, he can finish his typing in 3 hours. How long would it take him if he could type at 60 wpm?

   a) 4 hours b) 2 hours 15 min c) 3.5 hours d) 2 hours 40 minutes

3. How many minutes are in h hours?

   a) h/60 b) 60 h c) 60 + h d) 60 – h

4. If m pens cost d dollars, how many pens can you buy for w dollars?

5. A gear 50 cm in diameter turns a smaller gear 30 cm in diameter. If the larger gear makes 15 revolutions, how many revolutions will the smaller gear make in the same amount of time?

   a) 9 b) 12 c) 25 d) 15 e) 20

6. 10 yellow and 8 green balls are in an urn. Chad randomly draws two balls without replacement. If the first one he draws is yellow, what is the probability that the second one is also yellow?

MAKING COMPARISONS

Many times during your exam you will probably need to compare two numbers or to write a group of numbers in ascending (smallest first) or descending (largest first) order.

For any pair of numbers, one of the following relations must be true:

$a > b$ (a is greater than b)

$a < b$ (a is less than b)

$a = b$ (a is equal to b)

(Remember, the point always goes to the smaller number.)

To find the correct relation —**SHOOT THE PRODUCTS UP**

1. Write the 2 numbers as ratios with positive denominators
   (e.g. 3 can be written as $\frac{3}{1}$; $\frac{4}{-5}$ can be written as $\frac{-4}{5}$)
2. Put the 2 ratios in proportion form but leave out the = symbol.
3. Shoot the product of the terms in the **Extreme** position over the ratio on the left.
4. Shoot the product of the terms in the **Mean** position over the ratio on the right.
5. Whichever product is larger (**Extremes** or **Means**), so is the ratio above it.

WHAT'S LARGER $\frac{3}{7}$ or $\frac{9}{20}$?

ANSWER $\frac{3}{7} \times \frac{9}{20}$ since 63 > 60, $\frac{9}{20}$ is larger.

TRY THESE

1. Find the number that is less than $\frac{3}{4}$:

   a) $\frac{9}{10}$ b) $\frac{7}{9}$ c) $\frac{8}{11}$ d) $\frac{9}{12}$

2. Which is smaller: $\frac{-2}{3}$ or $\frac{-3}{4}$?

3. Which is larger: $\frac{3}{\sqrt{5}}$ or $\frac{\sqrt{5}}{3}$? [$\sqrt{5} \cdot \sqrt{5} = \sqrt{25} = 5$, Concept 6]

4. Arrange in order from largest to smallest $\frac{3}{7}, \frac{5}{12}, \frac{4}{9}, \frac{7}{15}$

PERCENTS

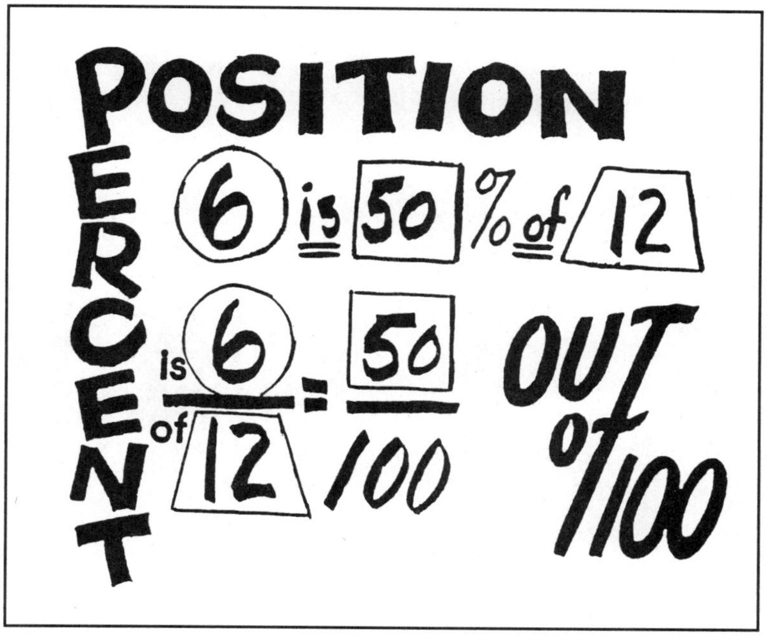

A percent is a ratio that compares a quantity to 100 of that same quantity. [*Whenever thoust sees % thoust thinks **100***] A tax of 8% means you pay the government 8 cents for every 100 cents you spend. A discount is what you don't pay. A **discount** of 30% means you pay 70% of the original or retail price. [If you know the discount subtract it from 100% to find the cost]. **A percent increase or decrease** compares the actual increase or decrease to the original amount. You can always solve problems involving percents by setting up a proportion. Just think: **"Is over of" is at 100.**

$$\frac{\text{IS}}{\text{OF}} = \frac{\text{AT(tached)}}{100}$$

QUESTION 35% of what number is 245?

ANSWER

1. Get ready for a proportion: $\frac{—}{—} = \frac{—}{—}$
2. Place 100 in bottom right for the % symbol and the number attached to the % on top. $— = \frac{35}{100}$
3. The term closest to IS goes on top left, the one closest to OF on bottom left. $\frac{245}{?} = \frac{35}{100}$
4. Reduce and solve. You may need to rewrite the problem using the words IS and OF. $\frac{7}{?} = \frac{1}{100}$

$? = 700$

TRY THESE

1. If 15 kilograms of pure water is added to 10 kilograms of pure alcohol, what percent of the resulting solution is alcohol?

   a) 66.6% b) 40% c) 25% d) 150% e) 10%

2. Charles weighed 200 pounds before starting his fitness program. He now weighs 160 pounds. What per cent of his weight did he loose?

3. A dress costs $12 after a 40% discount. What was the original price of the dress?

4. For which of the five products - A, B, C, D, and E, in the graph, was the percent increase in sales from 1987 to 1988 the same as the percent decrease from 1988 to 1989?

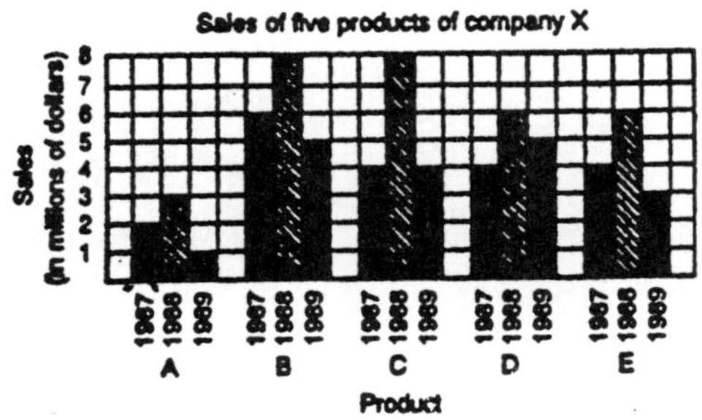

5. A 2 cup mixture is 1/3 flour and 2/3 cornmeal. If 1 cup of flour is added to make a 3 cup mixture, approximately what percent of the 3 cup mixture is flour?

   a) 65% b) 55% c) 50% d) 45% e) 35%

6. Maria was making $500 a week in May. In July she received a 10% increase. The following December, she received a 10% decrease. How much money was she making in December?

   a) $500 b) $495 c) $505 d) $540 e) none of the above

SQUARE ROOTS

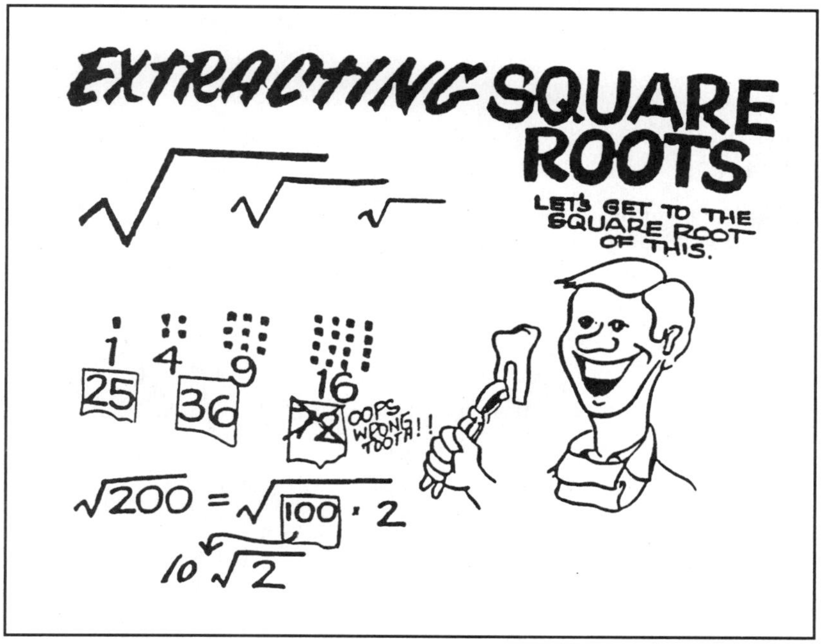

Writing numbers larger than 1 as square roots dramatically magnifies the appearance of the number. For example:

4 can be written as $\sqrt{16}$, 20 as $\sqrt{400}$

The square root of x^2 (written $\sqrt{x^2}$), is a number that when multiplied by itself is x^2.

$\sqrt{121} = 11$ or -11 $\qquad\qquad$ $\sqrt{y^{10}} = \pm y^5$

Perfect Square numbers include: **0, 1, 4, 9, 16, 25, 36, 49, 64, 81, 100, . . .**
These numbers have rational roots: **0, 1, 2, 3, 4, 5, 6, 7, 8, 9, 10, . . .**
Other integers such as: 2, 3, 5, 6, 7, . . . are not perfect squares and they have irrational roots. We can approximate their roots to any number of decimal places.

Note: When we take the square root of a number between 0 and 1 (say, 0.25), the root is larger than the number inside the radical. $\sqrt{0.25} = 0.5$

An interesting set of square numbers can be found when you square a number that ends in 5.

OBSERVE

$(45)^2 = 2025$ $\qquad$ $(35)^2 = 1225$ $\qquad$ $(75)^2 = 5625$

QUESTION $(95)^2 = ?$

ANSWER Observe that when we square a number that ends in 5 [such as 75] this leads to a result that ends in 25 [5625]. What proceeds the 25 [56] is the product of the number that proceeds the 5 [7] multiplied by the next integer consecutive integer [8]. Thus,

$(95)^2 = (9 \times 10)$ front + 25 back = $\underline{9025}$ and $(105)^2 = \underline{11{,}025}$

A square root is simplified, when all perfect square factors are removed from the term inside. To do this, pull out all perfect square factors, [the way a dentist pulls out a tooth] and write the root outside and leave the non-perfect square factors inside. You can also think of the number inside a radical sign as a PRISON SENTENCE. Imagine that a person gets 200 years in prison because he/she is part of organized crime. If he/she goes **square** and has **square** fact(or)s that can help nail worse criminals, they get a new identity [the square root] on the outside and a reduced sentence inside.

$\sqrt{200\pi}$ =	$\sqrt{100 \times 2\pi}$		=	10	$\sqrt{2\pi}$
prison sentence	**square fact**	**non-square**		**new identity**	**reduced sentence**

SIMPLIFY $\sqrt{0.25\pi}$

ANSWER $0.5\sqrt{\pi}$

TRY THESE

Simplify:

1. $\sqrt{75}$
2. $\sqrt{9x^2y}$
3. $\sqrt{5625}$
4. $\sqrt{0.0001}$
5. $\sqrt{m^{100}}$
6. $\sqrt{169k^{11}}$
7. Which of the following is not necessarily a perfect square, if a and b *are* perfect square integers?

   a) $25ab$ b) ab c) a^2b d) $a + b$ e) a^3b

SOLVING EQUATIONS AND INEQUALITIES

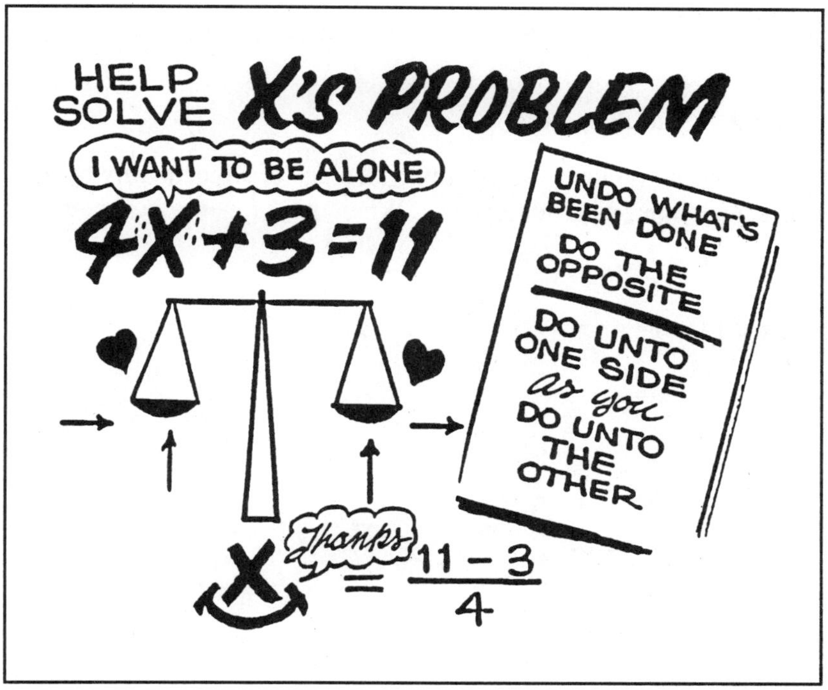

An equation is a sentence with an equals sign. It can be thought of as a balanced scale. The object of solving an equation is to make a variable **HAPPY**. A HAPPY variable is all alone on 1 side of an equals sign. Using the following two commandments of mathematics, it is easy to obtain a HAPPY variable.

I. **Undo what has been done by doing the opposite operation in the opposite order.**

For example: the expression 3x + 8 finds x having been multiplied by 3 then 8 being added on. Remember the order of operations? To UNDO this we will have to first subtract 8 (or add –8), then divide by 3 (multiply by $\frac{1}{3}$). Last hired, first fired.

II. **Do unto one side of an equation as you would do unto the other.**

SOLVE $3x + 8 = 20$

ANSWER

$$
\begin{aligned}
3x + 8 &= 20 \\
\underline{-8} \quad & \quad \underline{-8} && \text{subtract 8 from both sides obtaining} \\
\frac{3x}{3} &= \frac{12}{3} && \text{dividing both sides by 3 gives us} \\
x &= 4
\end{aligned}
$$

A statement with a $<$ or $>$ symbol in it is called in inequality. Inequalities may be solved using the same 2 rules as above.

if $3x + 8 < 20$ then $x < 4$

The **only** difference comes when you need to multiply both sides of an inequality by a negative number. That is, a negative number is squishing or squashing poor x. When this happens, a change in direction is needed.

EXAMPLE if $-3x < 12$ then $x > -4$

TRY THESE

1. Find J if $5J + 10J - 6 = 24$

2. If $3M + 11 = 10$, what does 6M equal?

3. What is one possible value of x for which

$$\frac{1}{5} < x < \frac{1}{4}$$

4. If $RS + P = 6P$
   and $P = -1$, find RS.

5. If a \$ b = ab + a and
   if 4 \$ 6 = x + 5, then x =

   a) 3 b) $\frac{14}{3}$ c) $\frac{24}{5}$ d) 23 e) 6

6. According to the formula: F = (9/5)C + 32, if the Fahrenheit (F) temperature was to increase by 27 degrees, how much would the Celsius (C) temperature be increased? [Hint: At 0° C, F = 32°, find C when F = 27° + 32° or 59°]

   a) 12 b) 15 c) 25 d) 48.6 e) 51

7. If the sum of y and y + 1 is greater than 7, but less than 11, which could be a value for y?

   a) –4 b) 2.4 c) 1 d) 3 e) 4

8. How many integers, n, are there such that:

   $$1 < 5n + 5 < 25$$

   a) Five b) Four c) Three d) Two e) One

Remember: **INT**egers are po**INT**s you usually fill in on a number line: [. . . –3, –2, –1, 0, 1, 2, 3, . . .]

Equations say the "darndest" things!

The equation: **2N + 3 = 13** when read left to right translates to:

"two times some number, N, plus 3 equals 13."

However, this also can be translated into:

(a) 13 is 3 more than twice N [reading right to left] or
(b) When 3 is added to twice a number the result is 13. [reading from middle out]

Write an equation that states each of the following, and then solve the equation:

9. When 7 is added to three times a number the result is 16.

10. 4 less than twice a number is 10.

THE DISTRIBUTIVE PROPERTY - PRAIRIE'S DEED + OTHER EASY WAYS TO COMPUTE

Multiplication is distributive over addition and subtraction. This means that you may change the order of operations to evaluate expressions where a number or expression appears right outside a grouping symbol (indicating multiplication).

Think of the numbers (or terms) inside the grouping symbol as male puppets on a children's show, and the number or term outside as *Prairie Dame*, a female puppet on the same show. Since there are generally so many more male puppets than female puppets, *Prairie Dame* has agreed to spend her TIME with all the male puppets. She will break the Girl Scouts' rule which says: evaluate inside grouping symbols first.

You will be able to multiply in your head quickly when you remember *Prairie's Deed.*

$$57(101) = 57(100 + 1) = 5700 + 57 = 5757$$

$$(75)(99) = 75(100 - 1) = 7500 - 75 = 7425$$

$$(8\frac{1}{2})(6) = 6(8 + \frac{1}{2}) = 48 + 3 = 51$$

It is important that you remember to distribute to ALL terms, especially when dealing with variables. Thus:

$$(a + b)^2 = (\mathbf{a} + \mathbf{b})(a + b) = (\mathbf{a} + \mathbf{b})(a) + (\mathbf{a} + \mathbf{b})(b) = a^2 + 2ab + b^2$$

QUESTION Which is larger: a) $3^5 + 3^6$ or b) 3^7?

ANSWER If we change each term to 3^5 times a number, we get:

a) $1 \times 3^5 + 3 \times 3^5$ and b) $3^2 \times 3^5$

which, gives us 4×3^5 compared to 9×3^5 and tells us that b is larger.

QUESTION Which is larger? a) $m^5 + m^6$ or b) m^7

Which is larger? a) $m^5 \times m^6$ or b) m^{11}

ANSWER If you Substitute 0 for m and then 1, –1, and .5, respectively for m, you will find that there is insufficient information to answer this question.

This is an identity so the two quantities are equal

TRY THESE

1. $x(x + y) + y(y - x)$ is equivalent to:

   a) $x^2 + 2xy - y^2$ b) $x^2 + y^2$ c) $x^2 - y^2$ d) $x + 3y$

2. Evaluate: $\left(2\frac{1}{2}\right)(14)$

3. A board 7 feet and 9 inches in length is cut into three equal parts. Find the length of each part. [7 ft. 9 in. = 6 ft. + 12 in. + 9 in.]

   a) 2 ft. 7 in. b) 2 ft. $6\frac{1}{3}$ in. c) 2 ft. 8 in. d) 3 ft. 7 in.

4. If $x + 3 = 10$, then $5x + 15 =$

   a) 30 b) 50 c) 70 d) cannot be determined

5. What's bigger: $2^{15} + 2^{16}$ or 2^{17}?

6. If $3x - 5y = 1$, which quantity is larger?

   A) $(-1)^{6x - 10y}$ or B) $(-1)^{9x - 15y}$

7. What is the value of $x^2 - y^2$ if $16 \div (x - y) = x + y$

There are easy ways to find out "how many" of something you have without actually counting. Simply break the problem down to one event at a time. When you say the word "and" it means multiply, when you say the word "or" it means add. Counting formulas also apply to questions asking you to find the probability of an event happening. For example, if the probability of taking route 101 is 0.6 and the probability of taking route 54, assuming that you did take 101, is 0.4 then the probability of taking route 101 and route 54 is 0.24

QUESTION How many four letter arrangements such as DCAE can be formed using the letters: A, B, C, D, and, E, if the first letter must be D, A must appear in the arrangement, and no letter can be used more than once in an arrangement.

ANSWER There are four slots to fill. The first one can be filled only one way, with a D, and the second is A, then the last 2 slots can be filled in 3 ways **and** the fourth in 2 ways (once you chose the other slots), **or** A can be in the third **or** fourth slot, so we have:

$$\underset{D\ A\ \_\ \_}{1 \times 1 \times 3 \times 2} + \underset{D\ \_\ A\ \_}{1 \times 3 \times 1 \times 2} + \underset{D\ \_\ \_\ A}{1 \times 3 \times 2 \times 1} = 18 \text{ arrangements}$$

There are easy ways to add consecutive integers. To add the first 100 integers, write the numbers forwards and backwards, add the columns, multiply sum by number of columns, and then divide the result by 2:

Sum of first 100 numbers:

$$\begin{array}{l} 1 + 2 + 3 + \ldots + 100 \\ \underline{100 + 99 + 98 + \ldots + 1} \\ 101 + 101 + 101 + \ldots + 101 \\ (100)(101)/2 = 50(101) = 5050 \end{array}$$

QUESTION If k denotes the sum of integers from 1 to 30 inclusive and m denotes the sum of the integers from 31 to 60 inclusive, what is the value of $m - k$?

ANSWER The sum of the first 1 to 30 integers is $30(31)/2 = 465 = n$. The sum of the integers from 31 to $60 = 91(30)/2 = 1365 = m$. Thus $m - n = 1365 - 465 = 900$.

TRY THESE

8. How many three-letter arrangements, such as ACE, can be formed using the letters: A, B, C, D, E, F, G, if the first letter must be B and one of the remaining letters must be E, and no letter can be used more than once in an arrangement?

   a) Five b) Six c) Nine d) Ten e) Twelve

9. There are 16 marbles in a bowl. Five are white, five are green, and the rest are yellow. Matt reached in the bowl and took one marble and then, without replacing the first marble, took another one. What is the probability that Matt picked two yellow marbles?

   a) 1/8 b) 9/64 c) 3/4 d) 1/16 e) 5/8

CONCEPT 9

AVERAGES OR MEAN AVERAGES

This concept will conclude our work in arithmetic and algebra. Traditionally when asked to solve a problem involving averages* (arithmetic means), you found the sum of a set of numbers and then divided by the number of entries. On Aptitude exams, you are often given an average (arithmetic mean) and the number of entries and need to find the sum. Since

$$\frac{\text{SUM}}{\text{ENTRIES}} = \frac{\text{AVERAGE}}{1}, \qquad \begin{array}{l}\text{SUM} = \text{ENTRIES} \times \text{AVERAGE} \\ \mathbf{S = EA}\end{array}$$

Once the sum is obtained, solving the problem will be easy. You will most likely be asked to find a missing entry and need to do some subtraction. Remember that if the average (mean) age of **10** people is **90**, the sum of their ages is equivalent to <u>each</u> person being 90 or $\mathbf{10 \times 90}$ which equals **900** years of experience to share. If you know the age of 9 of these people adds up to 800, then the missing age is 900 – 800 or 100 years old.

*Note: An average is the measure of the central tendency. There are 4 types of averages that begin with the letter **M**: mean, median, mode, and midrange. The median is the number in the middle when the numbers are arranged in order. The mode is the number that appears most often (it is the most fashionable), and the midrange is the arithmetic mean of the highest and lowest number. The Range is highest – lowest. For the test scores: 50, 50, 80, 90, 100 we find:

Mean: 74 Median: 80 Mode: 50 Midrange: 75 Range: 50

QUESTION If Josephine has a 97, 92, 100 and 93 on four exams, what must she get on her fifth exam to average 95?

ANSWER Josephine needs a sum of 95×5 or 475 to obtain a 95 average. So far, she has a total of 97 + 92 + 100 + 93 or 382. Thus she needs 475 – 382, or 93, on her last exam to obtain her desired result.

TRY THESE

1. Juan has a 79, 100, 92, 80, and 90 on five exams. What does he need on his sixth to average 83?

2. If the average weight of 5 coeds is 52 kg., and the coeds weigh 50, 60, 45, 55 and X, what is the value of X?

3. If Helene has a 94 average on 3 tests, what does she need on her 4th test to bring her average up to 95?

Questions 4 - 8 refer to the following table which shows the amounts collected from the first 25 donors in a charity drive.

Charity Drive Donations

Amount of Donation	\$1	\$5	\$10	\$20	\$25
Number of Donors	10	6	3	4	2

4. What is the mode of the amounts collected from the first 25 donors?

a) \$1 b) \$5 c) \$10 d) \$20 e) \$25

5. What is the median of the amounts collected from the first 25 donors?

a) \$1 b) \$5 c) \$10 d) \$20 e) \$25

6. What is the arithmetic mean of the amounts collected from the first 25 donors?

a) \$1 b) \$8 c) \$18 d) \$20 e) \$40

7. What is the midrange of the amounts collected from the first 25 donors?

a) \$13 b) \$8 c) \$12 d) \$12.5 e) \$24

8. What is the range of the amounts collected from the first 25 donors?

a) \$26 b) \$8 c) \$21 d) \$24 e) \$25

9. If the mean average of four numbers is 37, and the mean of two of them is 33, what is the mean average of the other two?

✓ Check Point 1

The questions below are very similar to questions asked on past exams.

1. If $x - 7 = 5 - x$ then $x =$

   a) –6 b) –1 c) 1 d) 6 e) 12

2. A recipe for a cake calls for $2\frac{1}{2}$ cups of milk and 3 cups of flour. With this recipe, a cake was baked using 14 cups of flour. How many cups of milk were required?

   a) $10\frac{1}{2}$ b) $10\frac{1}{4}$ c) $11\frac{1}{2}$ d) $11\frac{2}{3}$

3. $2x(3x + 5y) =$

   a) $5x + 7xy$ b) $6x + 10xy$ c) $6x^2 + 5xy$ d) $6x^2 + 10xy$

4. The cost of 30 melons is d dollars. At this rate, how many melons can you buy for 80¢?

   a) 24/d b) 240d c) 2.4d d) 3d/8 e) 8d/3

5. A class softball team won 2 games and lost 10. What part of its games did it win?

   a) 1/6 b) 1/5 c) 4/5 d) 5/6 e) 1/10

6. If m containers of milk cost d dollars, at this rate, what would s containers of milk cost?

   a) dm/s b) ds/m c) dms d) $d + m - s$ e) 1/msd

7. If a record album cost $7.20 after a 40% discount, what was the original price?

   a) $12 b) $7.60 c) $10.08 d) $11 e) $13

8. $33{,}333 + (4 \times 10^3) =$

   a) 34,333 b) 37,333 c) 33,433 d) 33,733 e) 333,700

For questions 9 through 16, answer:

A if the quantity in Column A is greater;
B if the quantity in Column B is greater;
C if the two quantities are equal;
D if the relationship cannot be determined from the information given

		Column A	Column B
___ 9.		$\frac{6}{15}$	$\frac{11}{30}$
___10.	If 25% of x = 60,	50% of x	120
___11.		$\frac{40}{9}$	4.5
___12.		The number of odd integers greater than 2 and less than 10	The number of prime integers greater than 2 and less than 10
___13.	The list price of a sweater is w dollars	The price of the sweater at 25% off the list price	0.75w
___14.	If m = 0.5	$m^5 + m^6$	m^7
___15.	If x + 3 = 5 and 2y + 11 = 15	x	y
___16.	If 2k + 3 < 9	k	3

17. If it takes 10 people 12 hours to do a certain job, how many hours would it take 6 people, working at the same rate, to do $\frac{1}{4}$ of the same job?

a) 6 b) 5 c) $4\frac{1}{2}$ d) 4 3) $3\frac{3}{4}$

18. If $2^n = 8$, what is the value of 3^{n+1}?

19. Which of the following is equal to $\frac{120 + n}{40}$?

a) $\frac{3 + n}{8}$ b) 3n c) 3 + n d) $3 + \frac{n}{40}$ e) 3 + 40n

20. In Thailand, a taxi cost $0.75 for the first kilometer, $0.50 per kilometer (or part thereof) for the next 5 kilometers, and $0.25 for each additional kilometer (or part thereof). If the charge for a fare was $4.50, which of the following could be the distance, in kilometers, of a taxi trip?

a) 9.5 b) 10 c) 10.5 d) 13 e) 17.5

21. If a bag has 48 balls that are either red, white, or green, and the probability of selecting a red one is 5/8, and white one is 1/8, how many green balls are there in the bag?

a) 6 b) 12 c) 16 d) 18 e) 24

22. If line AB intersects the y axis at (0, a), where a > 0, and the x axis at (2a, 0), what is the slope of line AB?

a) –2 b) –0.5 c) 0 d) 0.5 e) 2

23. Ms. Jones spent 1/4 of her savings on a car, and then 1/5 of the remaining amount on a computer. If she now has $3,000 left, how much was in her savings account before her purchases?

a) $4500 b) $5000 c) $5,500 d) $6,000 e) $7,500

24. 25 percent of 16 is equivalent to 1/2 of what number?

a) 2 b) 4 c) 8 d) 16 e) 32

25. If a team won 60% of the first 20 games it played, and 80% of the remaining 30 games in the season, what per cent of the total games played did it win?

26. If for all integers, P, $*P* = P^2$ if P is an even integer, and $*P* = P^2 - 1$ when P is an odd integer, what is the value of $*5* - *4*$?

a) 10 b) 9 c) 8 d) 1 e) 0

27. Jane bowled three games and averaged 120 points. Her highest score was 150. What was the average of her two lowest scores?

28. Mr. Smith's Calculus class has seven students. On the last test there were two 90's, one 85, one 80, and three 60's. Let M and m be the median and mean scores, respectively. What is the value of $M - m$?

EQUAL ANGLES

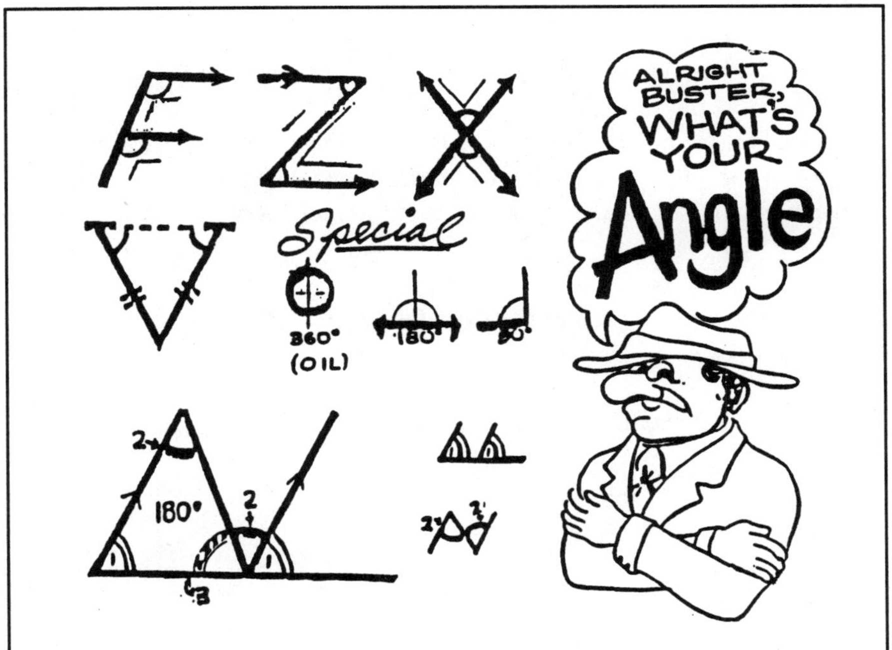

LET'S MEET AT THE OIL FIELD BETWEEN FV AND ZX

When solving problems that involve angles, it is often necessary to find pairs of equal angles on route to a solution. The four letters F, V, Z, X, in **any** position (technically called corresponding, base angles, alternate interior, and vertical) should be looked for. Think "Five Six", that is, "FiV ZiX." There are restrictions however. To use the "F" and "Z," you must know that parallel lines are present. The sides of the "V" must be two equal sides of a triangle. The other 2 angles in that triangle will be equal, and to use the "X," you must have straight lines that intersect.

Other letters to look for when dealing with angles are: "O" for 360°, "I" (or lazy l) for 180°, and "L" for 90°.

The Detective goes to the OIL field between five and six.

This information allows us to prove that in a plane triangle, the sum of all 3 angles is 180°.

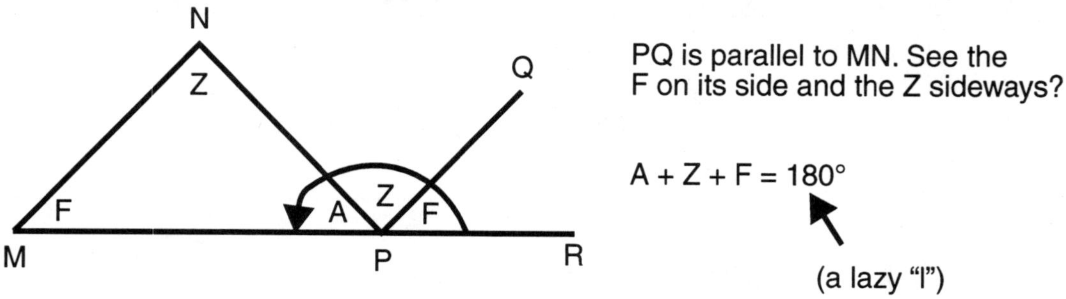

If you find a pair of angles that form the letters F or Z then the lines that look parallel will be parallel, and if you find two equal angles in a triangle then the opposite sides that make a V will be equal, and if you find two equal angles that form an X then their sides form two straight lines. [All of these angles are found on the same plane].

Note: In a triangle the largest angle lies opposite the largest side and the smallest angle lies opposite the smallest side.

Below you will find a convex quadrilateral, pentagon and octagon. The sum of the angles of a convex polygon of n sides is: (180)(n – 2)

$m\angle A + m\angle B + m\angle C + m\angle D = 360.$

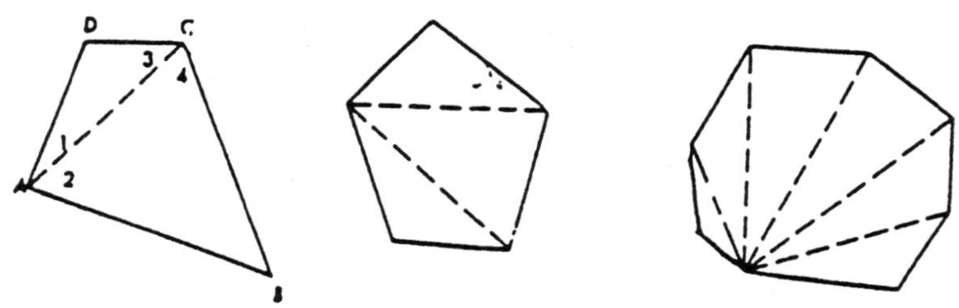

FIND Find t in the diagram below, given AC = BC and AE and BD are straight lines:

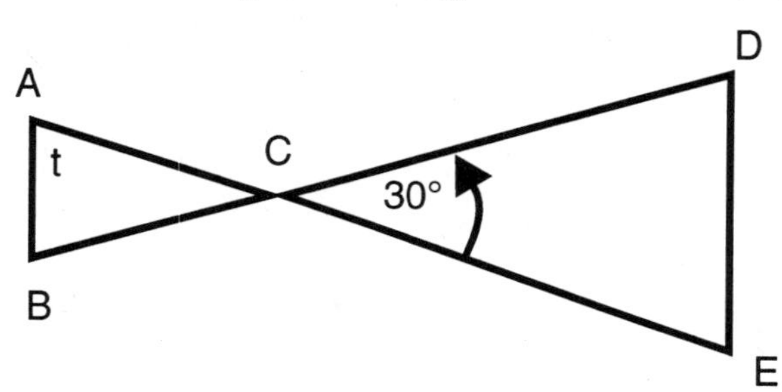

ANSWER Since AC = BC, we can locate a "V" (on its side) and know that the angles formed at A and B are both t°. Since AE and BD are straight lines, we know that the X angles, ACB and DCE are equal to 30°. Since every triangle has angle sum 180°, 2t + 30 = 180. Solving, we find t = 75°.

TRY THESE

1. If parallel lines L_1 and L_2 are intersected by L_3, what is the value of x + y?

a) 180° b) 150° c) 120° d) 90° e) can't be determined

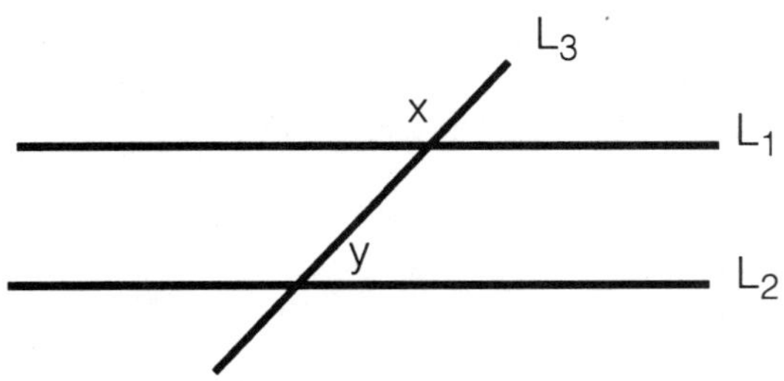

2. In the figure below, $L_1 \parallel L_2$ and AB = BC. If the measure of $\angle DBC$ is 53°, what is the measure of $\angle ABC$?

a) 53° b) 74° c) 90° d) 106° e) 127°

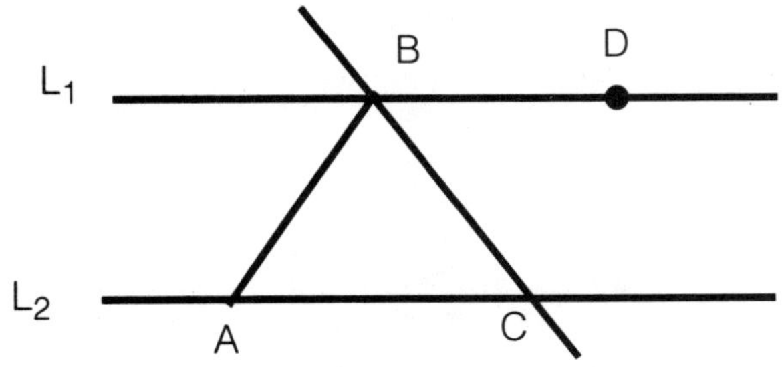

Note: Figure not drawn to scale.

3. If $\angle a = \angle b = \angle c$ and $\frac{d}{a} = \frac{2}{1}$ find d.

Line L is a straight line.

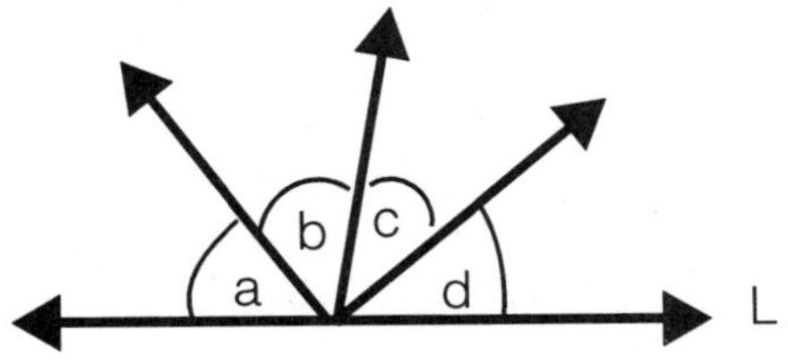

Note: Figure not drawn to scale.

4. If the sum of the angles of a convex polygon of n sides is 720 degrees then n =

a) 4 b) 6 c) 8 d) 360 e) 9

5. If AB = BX and XC = CD, $\angle BXC$ = a, and the other angles are as marked, what is a in terms of b and c?

a) 180 – 2(b + c) b) b + 180
c) b + c – 180 d) 2(b + c) – 180
e) –2b – c

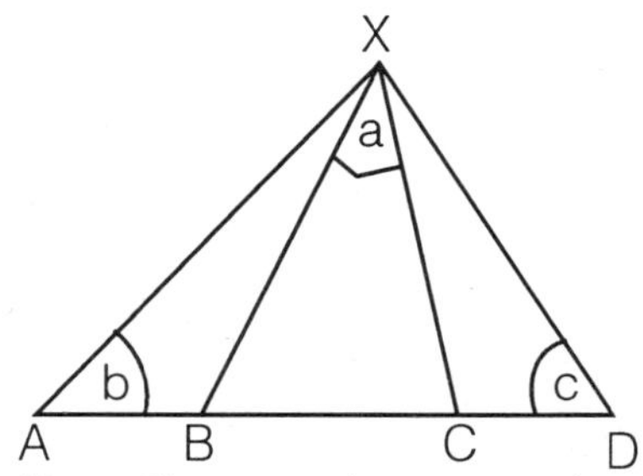

Note: Figure not drawn to scale.

6. In the figure on the right, find the value of angle x

a) 20 b) 40 c) 80
d) 160 e) 85

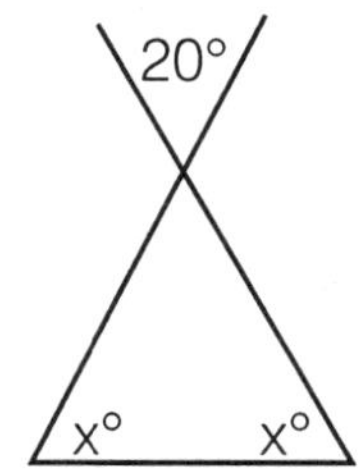

Note: Figure not drawn to scale.

7. In the figure on the right, AB is parallel to CD. What is the sum of the measures of the angles r + s?

a) 52 b) 142 c) 120
d) 128 e) 138

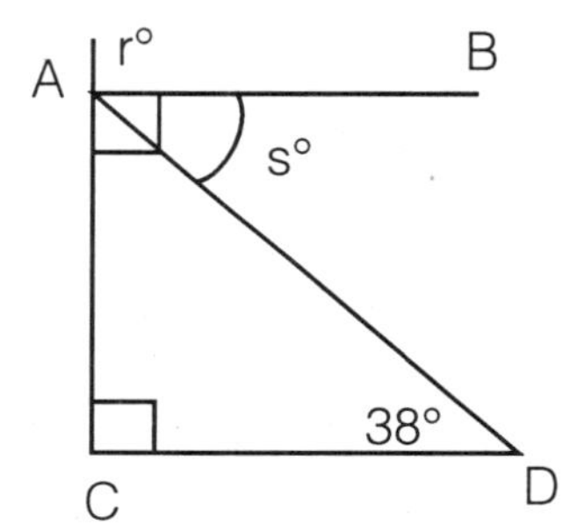

Note: Figure not drawn to scale.

AREA AND PERIMETER

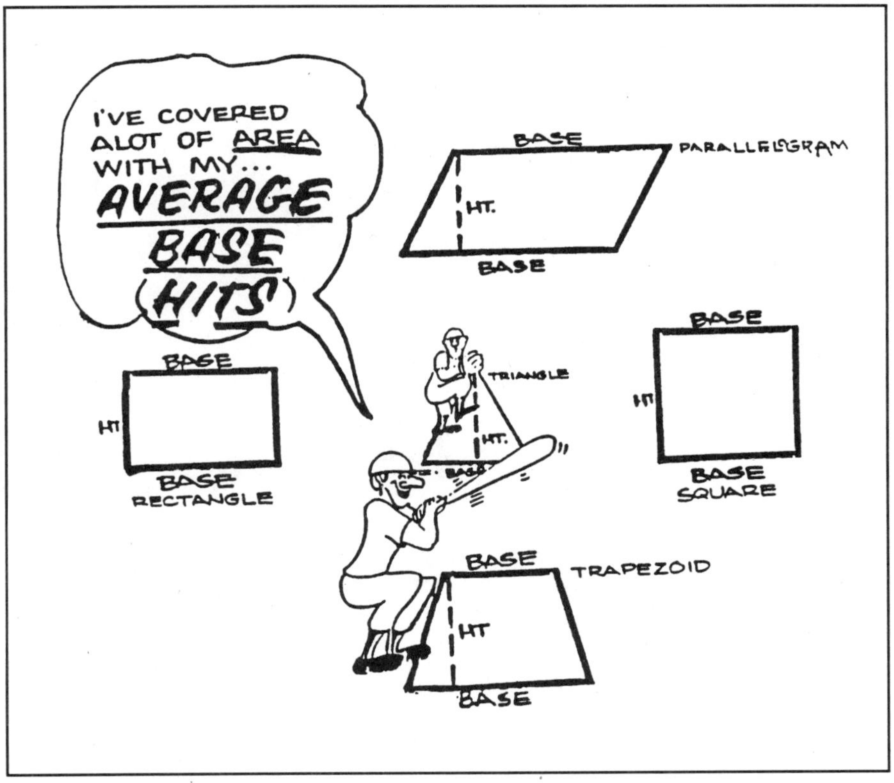

Finding the area and perimeter of geometric figures is a popular pastime on standardized exams. Area is the space inside a closed geometric figure. Pe**RIM**eter is the distance around the figure, or the sum of all sides.

Think of a baseball diamond where home plate is a trapezoid (a four-sided figure with 2 parallel sides or a triangle with its head cut off), first base is a square, second base a parallelogram (a four-sided figure with 2 pairs of parallel sides), third base a rectangle, and pitcher's mound a triangle. Imagine a prolific batter saying:

I've covered a lot of **area** with my **AVERAGE BASE HITs**.

With this picture in mind, we can find the area of all shapes mentioned.

Any side may be the **base** of a polygon. The side opposite and parallel to the base is called the "other **base**." The **height** is the distance between the two bases. (A triangle has one base, the other base is a point whose length is 0.)

To find the area, find the **average** of the **bases** and then multiply by the height. "**AVERAGE BASE HITs.**"

BASE (FIGURE)	FORMULA	SIMPLIFIED	EXAMPLE
Home Plate **Trapezoid**	Area = $\left(\frac{B_1 + B_2}{2}\right)h$	$\frac{1}{2}h(B_1 + B_2)$	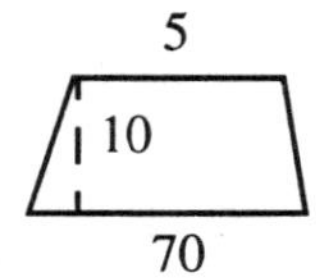 A = 70
First Base **Square**	Area = $\left(\frac{S + S}{2}\right)S$	S^2	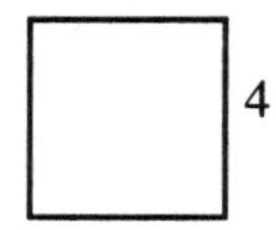 A = 16
Second Base **Parallelogram**	Area = $\left(\frac{B + B}{2}\right)h$	Bh	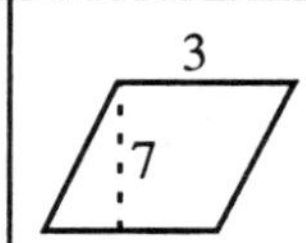 A = 21
Third Base **Rectangle**	Area = $\left(\frac{L + L}{2}\right)W$	LW	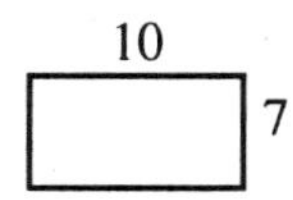 A = 70
Pitcher's Mound **Triangle**	Area = $\left(\frac{B + 0}{2}\right)h$	$\frac{1}{2}Bh$	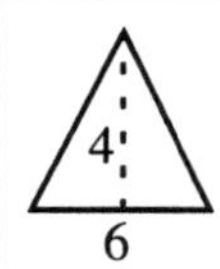 A = 12

TRY THESE

1. Find the area of the shaded region. ABCD is a rectangle.

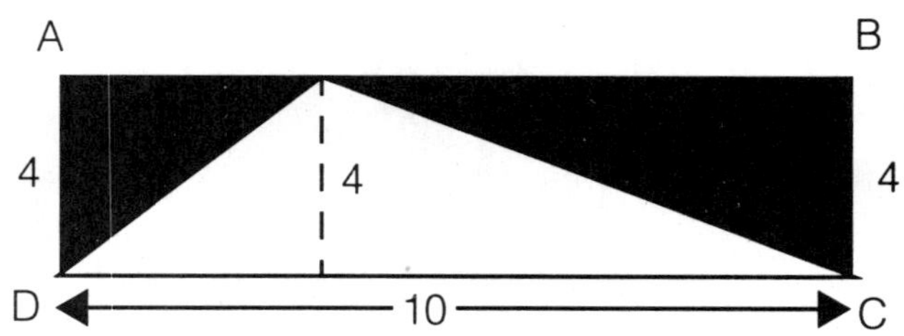

2. In the figure below, PQ is parallel to OR and OP is perpendicular to OR. What is the area of the trapezoid?

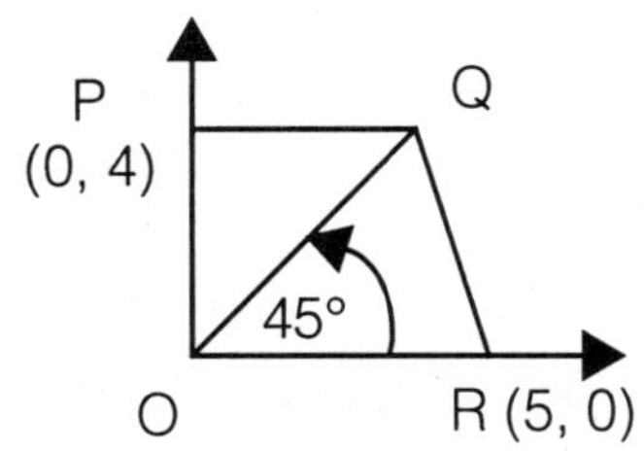

a) 9 b) 14 c) 18 d) 36 e) can't be determined

3. If the area of a square is 0.25, what is the perimeter of the square?

a) 0.0625 b) 1.00 c) 2.00 d) 2.25 e) 4.00

4. If the length of the sides of squares S and T are respectively $\sqrt{5}$ and $\sqrt{7}$, then the area of T minus the area of S is:

a) $\sqrt{2}$ b) 2 c) 4 d) –2 e) 24

5. A triangle has sides with lengths x – 2, 2x – 3, and x + 1. If the perimeter of the triangle is 20, what is the length of the longest side?

a) 9 b) 8 c) 7 d) 6 e) 5

6. The lengths of the sides of a triangle are x, 9, and 17, where x is the length of the shortest side. If the triangle is not isosceles, what is a possible value of x?

WHAT'S YOUR PREFERENCE?

If you had your choice, which method would you prefer to solve the following problem? Given: Parallelogram ABCD, DP is 1/3 of DC. The ratio of the area of triangle APD : ABCD is:

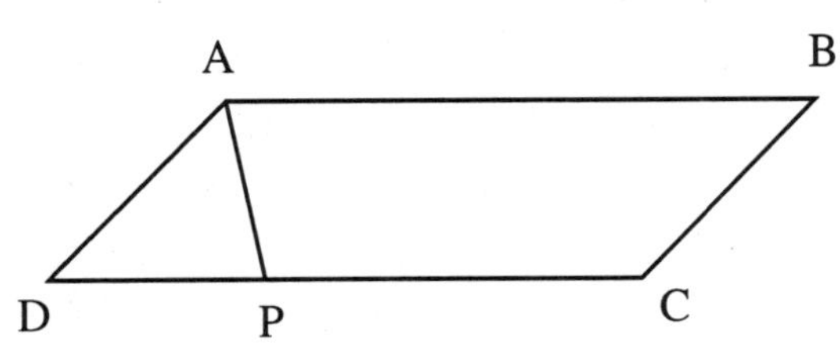

a) 1 : 3 b) 1 : 4 c) 1 : 6 d) 1 : 12

Method 1: Put in numbers

Since you are not told the measures of the height nor the base of either of these figures, and that information is necessary to solve the problem, make up your own numbers. Make them easy. Since you'll need to take 1/3 of the base of DC, find a number that has 3 as a factor, and make that the length of base DC. Let's use 6. This will make DP 1/3 of 6 or 6/3 or 2. Let's use 10 for the height of these two figures. Draw in the altitude or height from A to DC. The area of triangle ADP is (1/2)(2)(10) or 10. The area of parallelogram ABCD is (6)(10) or 60. The answer is . . . 10 : 60 or 1 : 6.

Method 2: Draw in line segments

In working with geometric figures, you are allowed to draw "auxilliary," that is, extra figures such as points and line segments to solve problems as long as you adhere to three rules: 1) The auxilliary figures must exist, 2) they must be unique, and 3) you cannot make any assumption about the properties of the figures unless it can be proven. For example, you can't draw in the bisector of an angle of a triangle and assume that it will be perpendicular to the opposite side.

Locate point Q on DC so that DP = PQ = QC. Locate points R and S on AB so that AR = RS = SB. Now draw PR, RQ, QS, and SC. Do you have 6 congruent triangles, that is, triangles the exact same size and shape? The ratio in question must be 1 : 6.

Method 3: Use Algebra

Let's call the length of DC, b and the height AH, h. DP must be 1/3b and the area of triangle APD is (1/2) (1/3) bh or (1/6)bh. The area of the parallelogram is bh. Thus, the ratio in question is 1/6 or 1 : 6.

Which of these methods works best for you? Ask a few friends or relatives which works best for them. Mathematics is a path from known information to what you want to know. There are many paths that you can take to accomplish your goals.

THE CIRCLE

The circumference of a circle is like the crust on a pizza. In every circle, the ratio of the circumference to the diameter is a real number that is a little more than 3. Mathematicians call this number pi (π) and usually approximate its value of 3.14 or 22/7. Thus if you know the radius or diameter of a circle, you can obtain the circumference by multiplying the diameter by π, or doubling the radius and then multiplying by π since C/D = π, C = Dπ. Think: **The Crust is de pie**.

When a wheel makes a complete revolution, 360°, it travels a distance equal to its circumference. A circle may be thought of as a regular polygon (a polygon all of whose sides and angles are equal) and infinitely many sides. The area of a regular polygon is 1/2 perimeter $\times$ radius of an inscribed circle. For example, the area of the hexagon is 1/2(10 + 10 + 10 + 10 + 10 + 10) $\times$ 8.5 = 255.

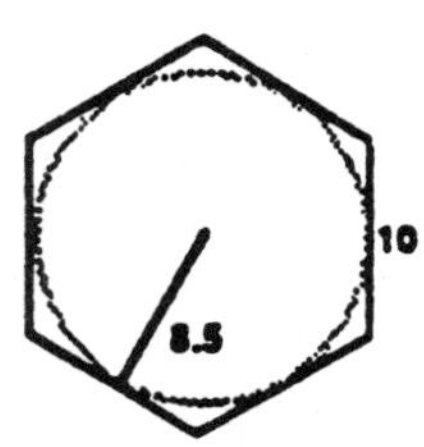

Since the circumference of the circle is the circle's perimeter, the area of a circle is equal to $1/2\,(2(\pi r) \times r)$. Simplifying, this gives us:

$$\text{Area} = \pi r^2 \text{ (pies are squared) or } \pi rr \text{ (pies are round)}$$

Thus, you can obtain the area of a circle if you know the radius of the circle; just square the radius and multiply by π.

QUESTION A square with area 64 has a semi-circle built on one of its sides. What is the area of the semi-circle?

ANSWER It is a good idea to draw a picture since the problem involves geometric figures. Thus we get:

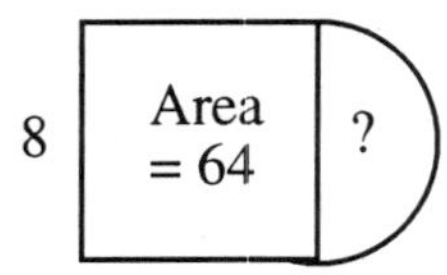

In order to find the area of the circle, we need to know the radius of the circle. From the diagram, we can see that the diameter is 8. Thus, the radius is 4 and the area of the entire circle is 16π. The problem, however, asks for the area of the semi-circle or half of the circle; thus, the answer to the problem is $1/2(16\pi)$ or 8π.

TRY THESE

1. If a triangle with base 10 is equal in area to a circle with radius 10, what is the height of the triangle?

   a) 10π b) 20π c) 100π d) 200π e) 1000π

2. How far does a wheel travel in 10 revolutions if the diameter of the wheel is 10 cm?

   a) 20π b) 100π c) 200π d) 10π e) 1000π

3. If you triple the diameter of a pizza, what happens to the crust? What happens to the area?

   a) triples, triples
   b) triples, increases by 9 times
   c) increases by 9 times, triples
   d) nothing, nothing

4. What is the perimeter of a square that has the same area as a circle whose circumference is π cm?

   a) $2\sqrt{\pi}$ b) 4π c) 0.5π d) 0.25π e) $0.5\sqrt{\pi}$

ANGLES AND CIRCLES

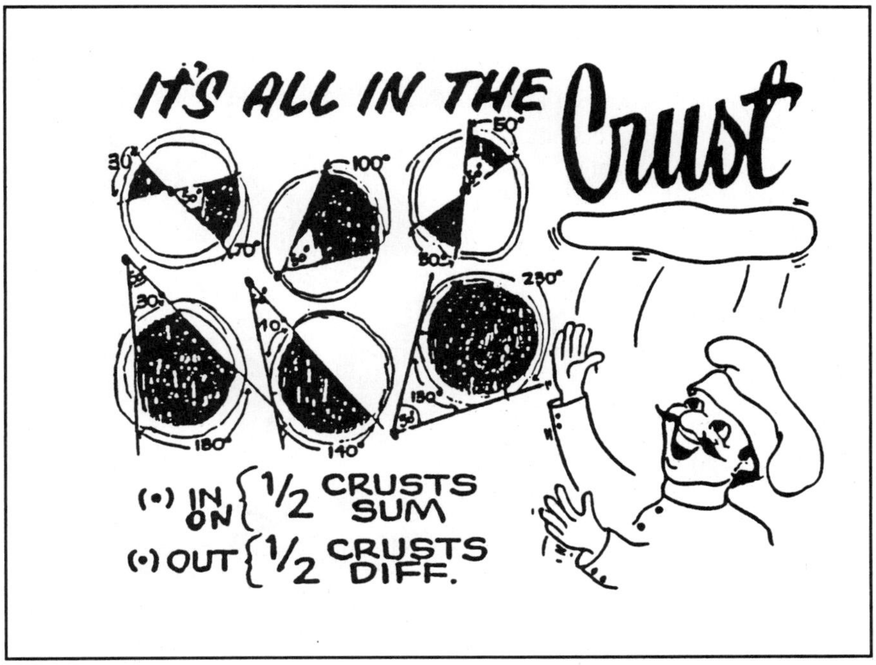

If an angle cuts out an arc (crust) of a circle, you can always obtain the measure of the angle if you know the measure of the crust. When the vertex (point) of the angle is the center of the circle, then the angle measure and the crust are the same. If the vertex is not at the center, then:

1. Locate the vertex position. Determine if it is inside, on, or outside the circle.
2. If the vertex is inside or on, then find the average of the arc it cuts out and the arc its vertical angle (X) cuts out. That is, the angle equals half of the sum of the **two** arcs.

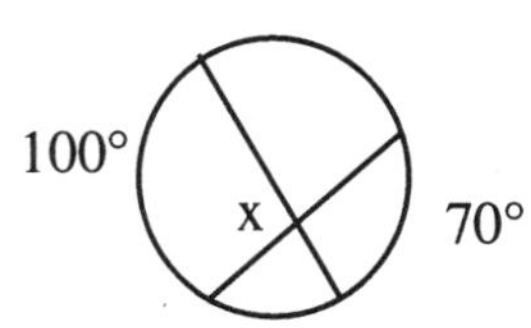

angle x = $\frac{1}{2}(170)$

3. If the vertex is outside, there's a DIFFERENCE. To find the measure, find half the difference between the two arcs that the angle will cut out of the circle.

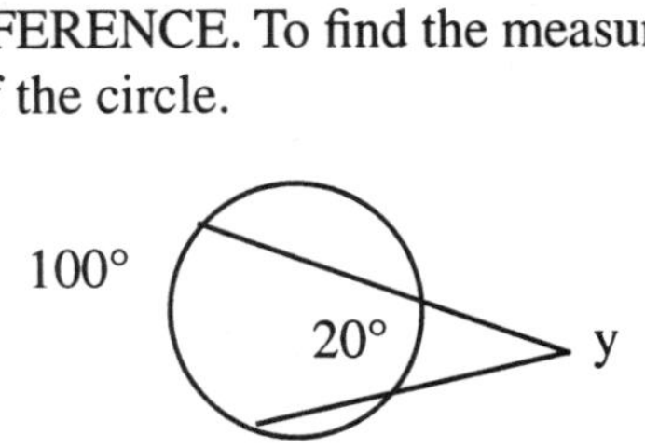

angle y = $\frac{1}{2}(100 - 20)$

FIND Find M + N in the diagram to the right.

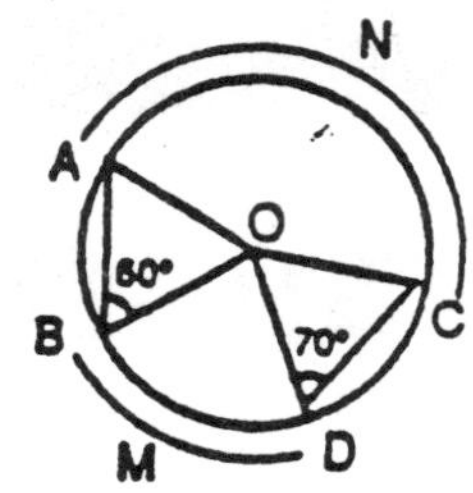

ANSWER Since this is a circle, the radii are equal. Thus, we can look for V angles. Since angle ABO is 60°, angle BAO is 60° and angle AOB must be 60°, also. Similarly, angle COD must be 40°, (180° – (70° + 70°)). Thus, arc $\overset{\frown}{AB}$ + $\overset{\frown}{CD}$ = 60° + 40° or 100°. This leaves 360 – 100 or 260° for M + N.

TRY THESE

☞ 1. $\overset{\frown}{AB}$ = 70° and $\overset{\frown}{DC}$ = 20°. Find x and y.

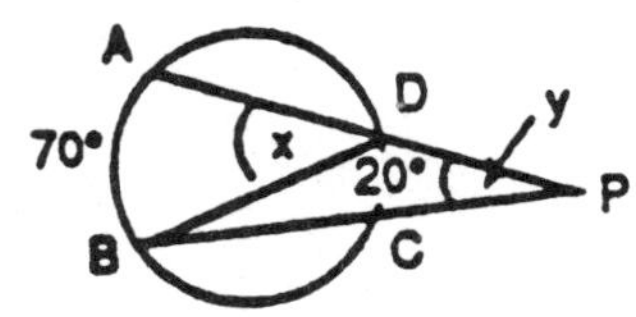

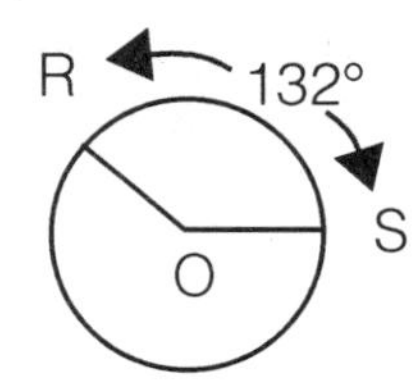

2. In a circle with center O, arc $\overset{\frown}{RS}$ (crust RS) = 132°. How many degrees in angle RSO?

   a) 18° b) 20° c) 26° d) 24° e) 48°

THE PYTHAGOREAN THEOREM

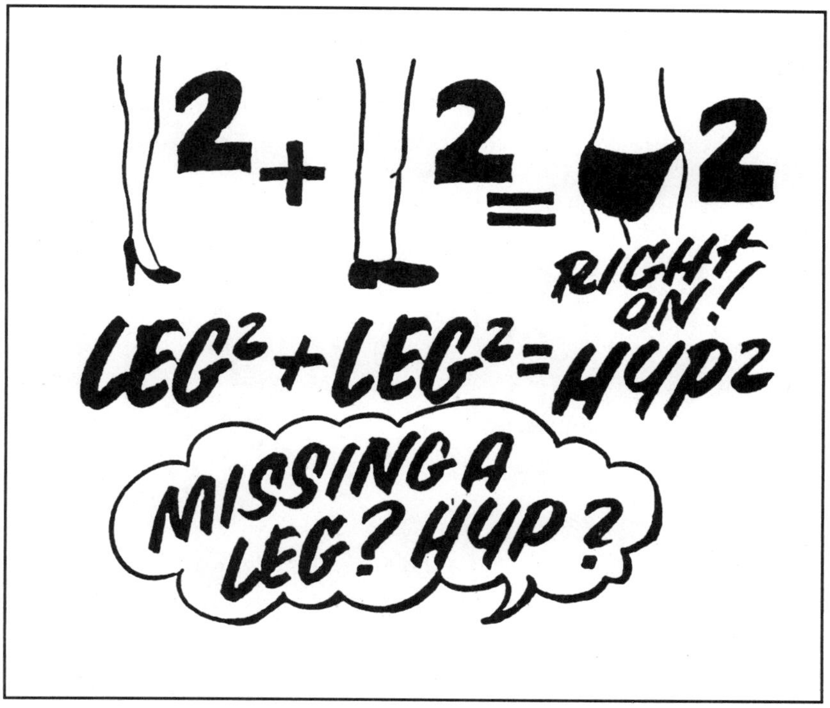

A Greek mathematician named Pythagoras gets credit for being the first person to prove what the ancient Egyptians knew, that in every right triangle, the sum of the squares of the legs (the sides that make up the right angle) is equal to the square of the hypotenuse (the third and largest side) of a right triangle.

There have been many original proofs of the Pythagorean Theorem since Pythagoras (a theorem is a statement that can be proven true). In fact, one of our presidents did an original proof of the Pythagorean Theorem. Guess who? (See answer page.)

If you know two sides of a right triangle, you can always find the missing side by using the Pythagorean Theorem. If two sides of a right triangle are in proportion, the third side will be in proportion as well. Applying this information can save you time on your exam. Look at the sets of numbers below. They are called Pythagorean triplets:

$$\text{Leg}^2 + \text{Leg}^2 = \text{Hyp}^2$$

L	L	H
3	4	5
5	12	13
8	15	17

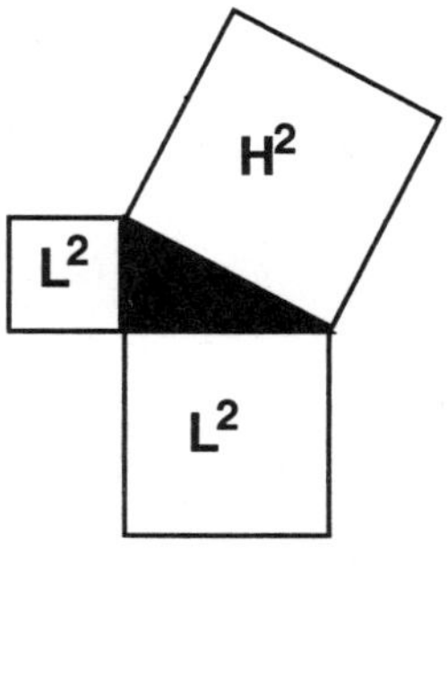

If you find a right triangle whose sides are proportional to two sides of a triplet, then the third side will automatically be proportional too. For example, if 2 legs of a right triangle are 30 and 40, to find the hypotenuse, go to the little Leg column and note that $10 \times 3 = 30$. Move across and note that $10 \times 4 = 40$. So moving to the hypotenuse column gives us 10×5 or 50 for the hypotenuse.

QUESTION If one leg of a right triangle is 36 and the hypotenuse is 39, what is the other leg?

ANSWER Checking our triplets, we find 39 is 3×13 and 36 is 3×12, thus our missing leg is 5×3 or 15.

TRY THESE

☞ 1. If two legs of a right triangle are 16 and 30, what is the hypotenuse?

2. Find the area of the shaded region.

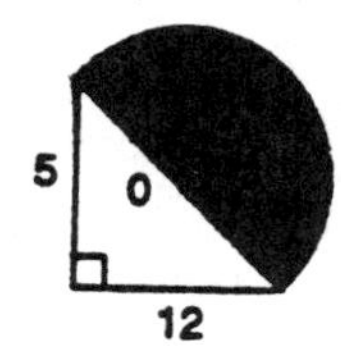

a) 169π b) $\frac{169\pi}{4}$ c) $\frac{169\pi}{8}$ d) $\frac{169\pi}{16}$ e) $\frac{13\pi}{2}$

Special Application of the Pythagorean Theorem

An equilateral triangle (pictured below) has all of its sides equal and all of its angles 60°.

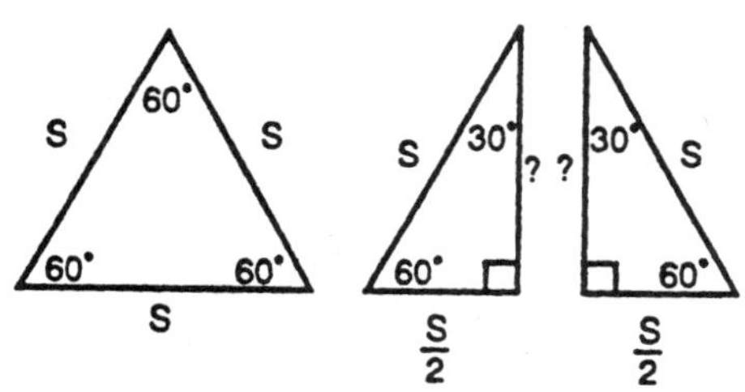

If you cut an equilateral triangle in half (see above), you will form two right triangles. Each of the new triangles has a 30° and 60° angle, a side from the original triangle, and a side that is half an original side. The missing side can be found by using the Pythagorean Theorem:

1. $\left[\frac{s}{2}\right]^2 + ?^2 = s^2$

2. $\frac{s^2}{4} + ?^2 = s^2$

3. $?^2 = s^2 - \frac{s^2}{4}$

4. $?^2 = \frac{4s^2 - s^2}{4}$ *(Concept 2)*

5. $?^2 = \frac{3s^2}{4}$

6. $? = \sqrt{\frac{3s^2}{4}}$ *(Concept 6)*

7. $? = \frac{s}{2}\sqrt{3}$

Remember: 30 is **half** of 60 and the smallest side is **half** of the largest side. The middle side equals the small side $\cdot \sqrt{3}$

In a 45°, 45°, 90° triangle, **two** sides and **two** angles are equal and the hypotenuse = leg $\sqrt{\text{two}}$ or leg $\cdot \sqrt{2}$

FIND Find the area of the semi-circle built on the right triangle.

ANSWER The diameter of the semi-circle is the hypotenuse of the right triangle. Since the two legs of the right triangle are equal, the hypotenuse equals $6\sqrt{2}$. The area of the semi-circle is:

$\frac{1}{2}\pi r^2$ or $\frac{1}{2}\pi(3\sqrt{2})^2$ or $\frac{1}{2}\pi(9)(2)$ $= 9\pi$ square units.

✓ Check Point 2

1. In two hours, the minute hand of a clock rotates through an angle of:

a) 60°   b) 90°   c) 130°   d) 360°   e) 720°

2. Two houses started out at the same price. When sold, house A increased by 20%, and house B decreased by 20%. What was the ratio of the selling price of A to B?

a) 6:5   b) 3:2   c) 2:3   d) 1:1   e) 12:11

3. In terms of the square units in the figure below, what is the area of the semi-circle?

a) 32π b) 16π c) 8π
d) 4π e) 2π

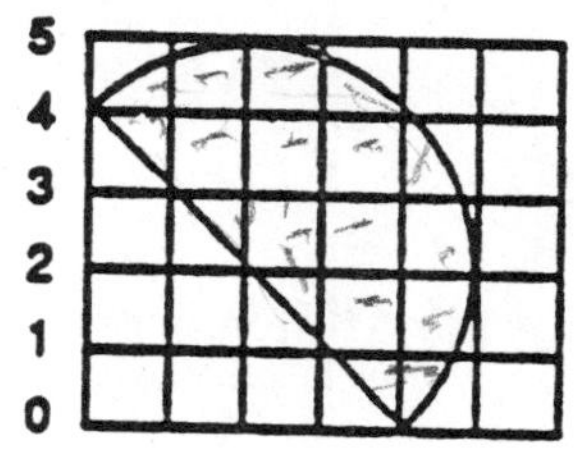

4. In circle O below, AB is a diameter. Angle BOD contains 15° and angle EOA contains 85°. Find the number of degrees in the angle ECA.

a) 15   b) 35   c) 50
d) 70   e) 85

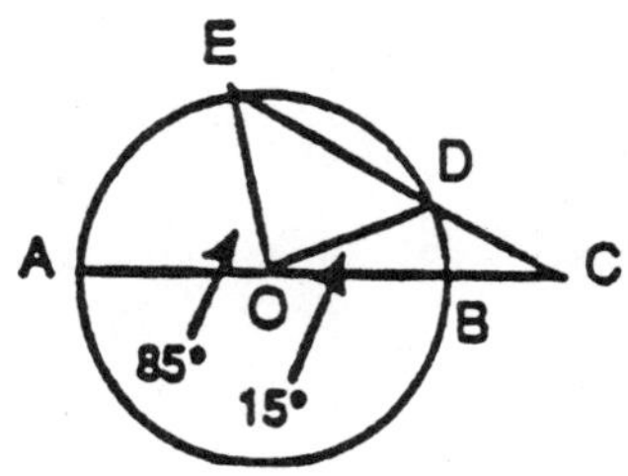

5. The average (arithmetic mean) of n numbers is 10 and the average of 5 of these numbers is 8. In terms of n, what is the average of the remaining numbers?

a) (10n + 40)/5   b) (10n – 40)/5
c) (10n – 40)/(n– 5)   d) (10n + 40)/(n – 5)

6. In the figure, PS is perpendicular to QR. If PQ = PR = 26 and PS = 24, then QR =

a) 14   b) 16   c) 18
d) 20   e) 22

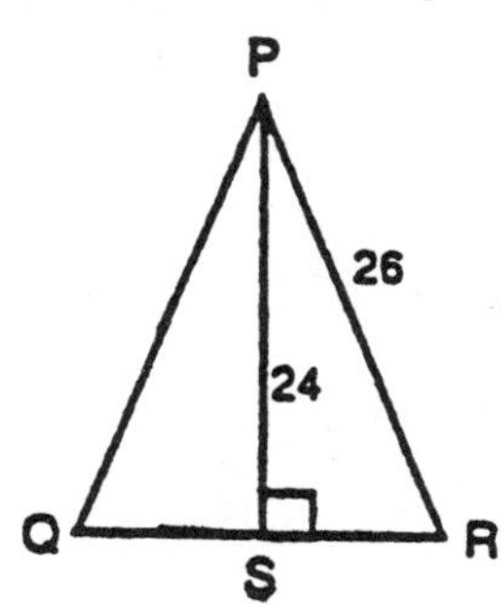

7. How many degrees between the hands of a clock at 3:40?
(Hint: 60 min. = a 30° move for the little hand.)

a) 150° b) 140° c) 130° d) 125° e) 120°

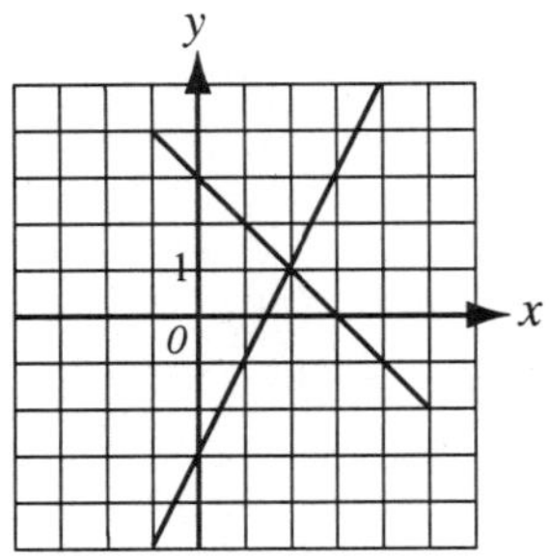

8. The equations $y - 2x = -3$ and $x + y = 3$ are graphed above. The solution for these simulataneous equations is:

a) $x = 0, y = -3$
b) $x = 0, y = 3$
c) $x = 1, y = 2$
d) $x = 2, y = 1$
e) $x = 3, y = 0$

9. If q and r are two different odd numbers and $q > r$, which of the following must also be odd?

I. $q + r$ II. $q - r$ III. $q r$

a) I only
b) III only
c) I and II only
d) II and III only
e) I, II, and III

10. If a face of a cube has area of 100 what is the volume of the cube?

a) 100 b) 400 c) 600 d) 1000 e) 2500

11. If $4/x = 8$, then what is $x - 1$?

a) −1.5 b) −2/3 c) −1/2 d) 1 e) 1/2

12. If there are 90 students in a school taking Physics and/or Chemistry and 50 are taking Chemistry, and 20 are taking both Physics and Chemistry, how many are taking only Physics?

a) 50 b) 40 c) 70 d) 60 e) 30

13. From the following choices how many outfits are possible, if an outfit consists of a shirt, tie and slacks

shirt - green, yellow, blue
tie - striped, plaid, solid
slacks - jeans, cords

a) 8 b) 9 c) 18 d) 16 e) 11

14. A rectangular garden that is 6 meters by 8 meters, is surrounded by a walk that is 2 meters wide. What is the area of the walkway?

a) $120\ m^2$ b) $72\ m^2$ c) $27\ m^2$ d) $48\ m^2$ e) $32\ m^2$

THE OTHER 10 PERCENT

Here are some concepts that appear on Aptitude tests from time to time. To get the extra edge, check these concepts out:

Forming Outfits, Forming Lines, Making Arrangements, etc.

If Jane has 10 shirts and 6 skirts, then she can form 10×6 or 60 different shirt-skirt outfits. Each shirt can be worn with 6 different skirts; $10 \times 6 = 60$.

In general, if an event A can occur in **a** ways, and an event B can occur in **b** ways, then events A and B can occur in **$a \times b$ ways.**

If Joe, May, Kate and Bob are forming a line, then there are 4 people that could come first, three second, 2 third and 1 fourth.

$$\underset{(1st)}{4} \times \underset{(2nd)}{3} \times \underset{(3rd)}{2} \times \underset{(4th)}{1} = 24$$

or 24 possible arrangements. (An arrangement where order counts is called a permutation.)

Solving Simultaneous Equations

The object of solving two equations with 2 variables is to form one equation with one variable and then apply Concept 7. Here are two easy ways to do this:

1. Addition or Subtraction
2. Substitution

Addition or Subtraction If adding one side of one equation to one side of another results in the elimination of one variable, this is a great method to use.

EXAMPLE If $x + y = 6$ and $x - y = 10$, find x and y.

ANSWER Lining up equations we get:

$$\begin{aligned} x + y &= 6 \\ x - y &= 10 \\ \hline 2x &= 16 \\ x &= 8 \end{aligned}$$

Adding the left sides of the equation eliminates y, thus by adding equals to equals we obtain **$x = 8$**.

To determine y, substitute $x = 8$ in either of the first two equations: $8 + y = 6$ or **$y = -2$**.

Substitution If one variable is solved in terms of the other (or can be easily put into this form), then this is the method to use. Simply substitute the expression the variable is equal to in the other equation. Then you will be ready to apply Concept 7.

EXAMPLE If $y = 2x + 5$ and $x + y = 20$, find x and y.

ANSWER Substituting $2x + 5$ for y in the second equation gives us:

$$\begin{aligned} x + y &= 20 \\ x + (2x + 5) &= 20 \\ 3x + 5 &= 20 \\ 3x &= 15, \text{ or } x = 5 \end{aligned}$$

Solving for y, we obtain $y = 15$.

Rules for Working with Exponents

1. $a^b a^c = a^{b+c}$
2. $a^b \div a^c = a^{b-c}$
3. $(a^b)^x = a^{bx}$

Make sure you have the same base number (a).

EXAMPLE If $x^7 = 3$ and $x^6 = \frac{2}{y}, y \neq 0;$

then $\frac{x^7}{x^6} = \frac{3}{\left(\frac{2}{y}\right)}$ and $x = \frac{3y}{2}$.

Types of Numbers

Two common ways of classifying whole numbers are:

- prime
- **com**posite

If you can think of more than one **com**bination of whole numbers that multiply to give you a number, then that number is **com**posite. For example, $12 = 3 \times 4$ or 1×12 or 6×2, so 12 is composite. The only whole number combination that will multiply to give you 17 is 1×17. 17 is **prime**. Other prime numbers include 2, 3, 19, 29, etc.

To check if a number is prime, it is sufficient to see if it can be divided, without a remainder, by 2, 3, ... up to the square root of the number.

Integers are often identified as **poin**ts on a number line: . . . –3, –2, –1, 0, 1, 2, 3, . . .

Rational numbers can be written as the **ratio** of two integers, p, q, where $q \neq 0$

Real numbers are the Rationals and the Irrationals such as π and $\sqrt{2}$

Probability

If there are 3 polka dot balls in a bag and 5 striped balls then the probability of picking a polka dotted one is the ratio:

$$\frac{3}{8} = \frac{\text{(polka dotted)}}{\text{total}}$$

The probability of an event is the ratio of favorable ways the event can occur to total number of things that can occur.

The odds of something happening is the ratio of favorable to unfavorable.

Volume

The measure of the space inside a three-dimensional region is called volume. It is measured in cubic units. The four most popular regions that you may need to find the volume of are:

Prisms and Cylinders

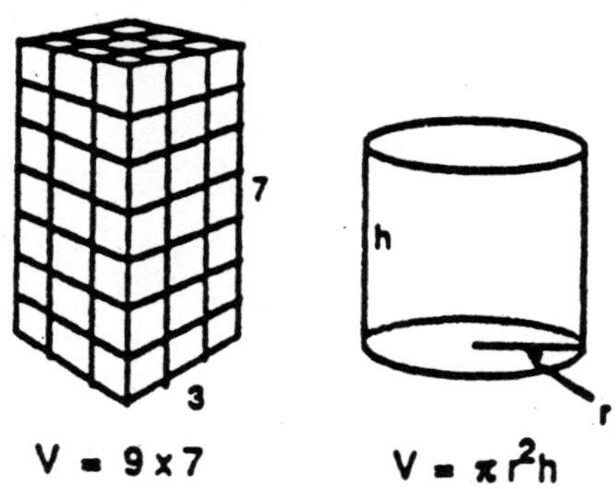

Volume = base area × altitude

Pyramids and Cones

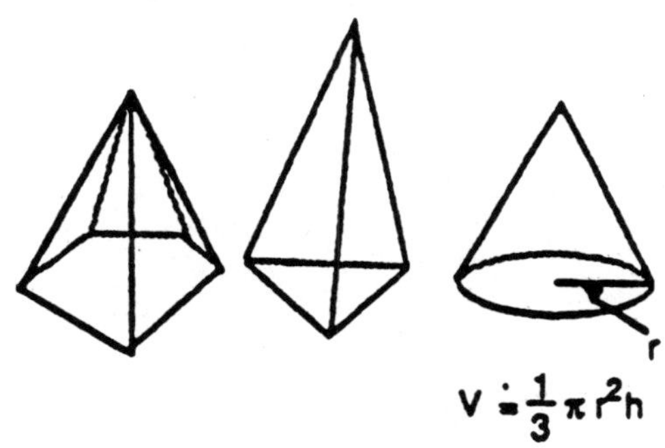

Volume = $\frac{1}{3}$(base area × altitude)

EXAMPLE Find the volume of the box formed by folding the figure on the right on the dotted lines.

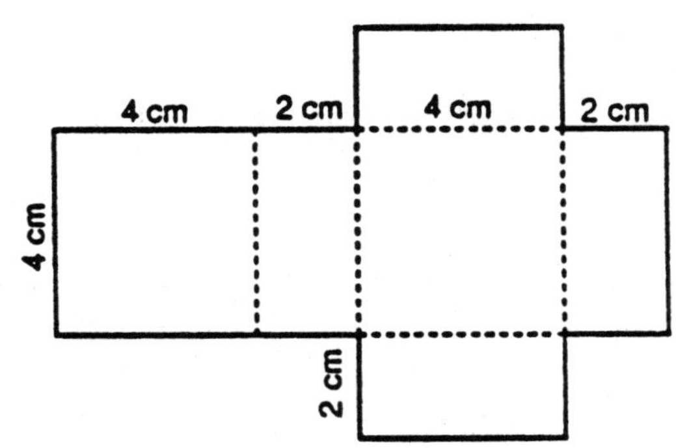

ANSWER The area of the base is 16, the altitude is 2, so the volume is given by base × altitude or $16 \times 2 = 32$.

INDIRECT METHODS OF SOLVING + GUESSING STRATEGIES

The first time through the test you should use your intuition and find the problems that are easiest for you to solve. The second time through, remind yourself of the 14 major concepts and look for key words to help you devise an effective strategy to solve the remaining problems. The third time, try an indirect method.

The Goldilocks Method: (a.k.a. guess and check)

When you read a problem with only similar looking numerical answers to choose from and it looks like a) something you vaguely remember from your Algebra class b) something that will take a long time to solve c) something you do not particularly enjoy working on or d) all of the above, then give GOLDILOCKS a try:

1. Arrange the answers in order from smallest to largest
2. Choose the median (middle) answer
3. Check that answer in the problem and decide if it is: too big, too small or just right. *[Thus, the Goldilocks method].*
4. If the answer is **not** "just right" then cross out that answer and two other answers that would give you a worse result.
5. Choose one of the remaining two answers. Check that choice and you will then know which answer is correct.

The Politicians Choice

When you look at the choice of answers and they do not all look the same, choose the one that looks the most like **all** the other answers. If the choices are all numerical and three look the same and two look "weird" try the median of the similar ones. If possible, check to see that you chose the correct answer.

TRY THESE

☞ 1. When 2/5 of a certain number is added to the number and the sum is subtracted from 108, the result is the number. What is the original number?

a) 35 b) 45 c) 54 d) 135/2 e) 270

2. Dan spent $9 on tomatoes and avocados to make a salad that required twice as many tomatoes as avocados. Tomatoes cost 10¢ each and avocados cost 25¢ each. How many avocados did he buy?

a) 40 b) 20 c) 25 d) 30 e) 10

3. A man was 28 years old when his son was born. How many years old will the son be when the father is 3 times as old as his son?

   a) 7   b) 14   c) 21   d) 28   e) 42

4. On the first day of a sale, 1/3 of the TV sets in stock were sold. After 2 more were sold, 3/5 of the original amount remained. How many sets were in stock?

   a) 10   b) 20   c) 30   d) 40   e) 50

5. How many liters of pure alcohol must be added to a 100 liter solution that is 20% alcohol to produce a solution that is 1/3 alcohol?

   a) 10   b) 40   c) 20   d) 10/3   e) 27

Making Up Numbers

Questions that require you to determine a ratio or make a generalization may be easily answered by making up numbers and working them out in the problem.

Make up a number that is easy to work with but don't use 1 or 0 because they have too many special properties. DO use 100 if you're dealing with percents. DO use 10 if you need to multiply in order to obtain a solution.

TRY THESE

☞ 1. If x – 3 is one greater than y, then x + 10 is how much greater than y – 2?

   a) 16   b) x – 5   c) 8   d) 32   e) x + y

2. The sum of three consecutive odd integers is N. In terms of N, what is the sum of the next three consecutive even integers that follows the greatest of the three odd integers?

   a) N + 6   b) N + 12   c) N + 15   d) N + 18   e) cannot be determined

3. If Sarah drives at 40 mph to her aunts house, and returns over the exact same route, at 60 mph, what was her average mph for the entire trip?

   a) 50   b) 45   c) 48   d) 58   e) none of these

4. If x is the average (arithmetic mean) of k and 10, and y is the average of k and 4, what is the average of x and y? (Make up a number for k.)

a) 14 b) 7 c) (k + 7)/4 d) (k + 7)/2 e) (k + 14)/2

5. If x + y = m, and x – y = p, then 2xy =

a) $(m^2 - p^2)/2$ b) $(p^2 - m^2)/2$ c) (m – p)/2
d) mp/2 e) $(m^2 + p^2)/2$

[Hint: If we let x = 5 and y = 2, then m = 7 and p = 3; so. . . . .2xy = 20. Which answer also equals 20?]

6. Last year Department Store X had a sales record for December that was 4 times the average (arithmetic mean) of the monthly sales totals for January through November. The sales total for December was what fraction of the sales total for the year?

a) 1/4 b) 4/15 c) 1/13 d) 4/11 e) 4/5

[Hint: If the average from Jan. to Nov. was 100, then Dec. would be: 400 and the total for the year would be: 1500, thus: compare December : years total]

MORE PROBLEMS TO PONDER — ANGLES AND CIRCLES

You will be given a few facts and a few measurements. From these you are to find the measure of 3 arcs and 20 angles. Geometry is a bridge from what you know to what you want to know.

FACTS

1. The sum of the angles of a triangle equals 180°.
2. The number of degrees in a circle, i.e., complete revolution, is 360°.
3. If the vertex of an angle lies on or inside of a circle then locate its vertical angle. Find its arc and its vertical angle's arc and take 1/2 of the sum of the two arcs (crust). This will be the measure of the angle (and the vertical angle).
4. If the vertex lies outside the circle then it will cut off two arcs. Take 1/2 the difference of the two arcs to find the measure of the angle.

GIVEN Circle O, diameter CM and Diameter ED.

The measure of arcs: $\overset{\frown}{CD} = 70°$, $\overset{\frown}{AC} = 20°$, $\overset{\frown}{EB} = 50°$.

FIND The measure of arcs:

1) $\overset{\frown}{AB}$ 2) $\overset{\frown}{EM}$ 3) $\overset{\frown}{DM}$

Find the measure of angles:

4)	COD	5)	EOM	6)	CMD
7)	BKC	8)	DKM	9)	DPM
10)	EMC	11)	EOC	12)	DOM
13)	BSC	14)	EMZ	15)	BMC
16)	BKM	17)	DEM	18)	BDO
19)	EDM	20)	BDM	21)	MDP
22)	BCS	23)	ABD		

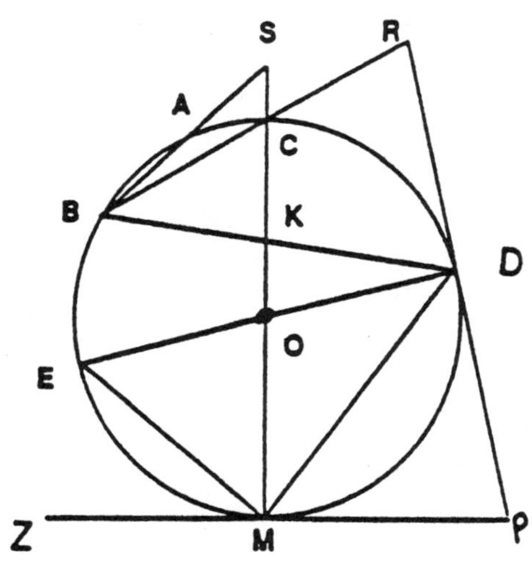

SUFFICIENT INFORMATION

In this section, you are given two numbered facts and asked to determine whether the information is sufficient to answer the question. There are five possible answers:

A) Statement 1 alone is sufficient
B) Statement 2 alone is sufficient
C) Both statements 1 and 2 together are sufficient
D) Either statement by itself is sufficient
E) Not enough information to answer the question

You are allowed about 1 minute and 15 seconds to answer each question. You do not have to solve each problem twice to obtain a solution. Here are some quidelines you can use to obtain a successful solution.

TOPIC	QUESTION	INFORMATION IN STATEMENTS 1 & 2	BEST CHOICE
Variables	Determine a ratio between 2 variables	Each statement gives a comparison between these two variables.	D
"	"	Both statements involve a 3rd variable.	C
"	Find the value of a variable	Each statement is an equation involving this and another variable. No other information is given.	C
"	"	Each statement is an equation involving this and another variable. Specific information is given about one or both of the variables.	A or B

TRY THESE

☞ ______ 1. How many glus = 8 mus?

1) 6 glus = 11 mus 2) 5 glus = 34 mus

______ 2. What is the value of y?

1) $y = 3x - 6$ 2) $x + y = 12$

______ 3. How many stubs in 8 dubs?

1) 8 dubs = 11 glubs 2) 12 glubs = 10 stubs

_______ 4. If x and y are whole numbers, greater than 0, what is the value of x?

1) $x + y = 2$ 2) $x - y = 5$

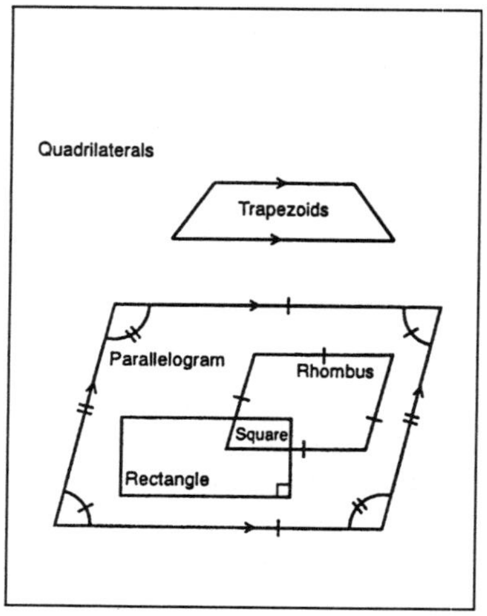

TOPIC	QUESTION	SUFFICIENT INFORMATION
Quadrilateral	Is this a parallelogram?	1) 2 pairs of parallel sides, or 2) 1 pair of parallel and equal sides, or 3) 2 pairs of equal sides
Parallelogram	Is this a rhombus?	One pair of equal adjacent sides
Parallelogram	Is this a rectangle?	One right angle
Rectangle	Is this a square?	One pair of equal adjacent sides
Rhombus	Is this a square?	One right angle
Two lines	Are these lines parallel?	Equal "F" or "Z" angles
Circle	Find area	The radius, diameter, or circumference
Logic	Is an element in a set?	Knowledge of exactly where element needs to be placed

_______ 5. Is ABCD a rhombus?

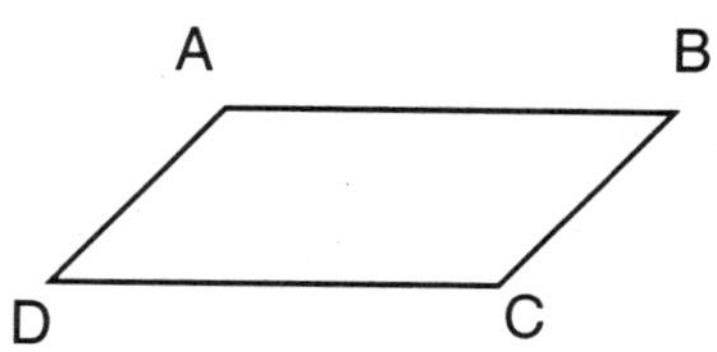

1) AB is parallel to DC.
2) AB is equal to DC.

_______ 6. Is line l parallel to line m?

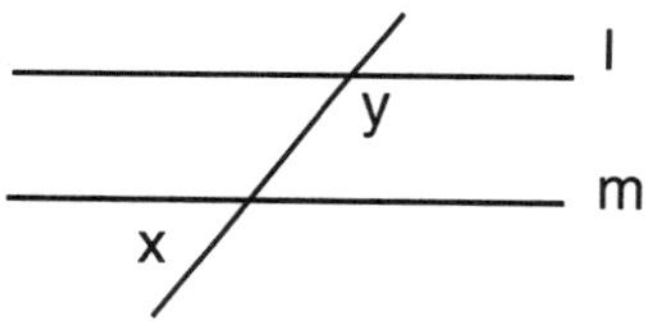

1) angle x = 70 degrees.
2) angle y = 110 degrees.

_______ 7. Does Jack live in Texas?

1) All of Sam's friends live in Texas.
2) Jack is Sam's friend.

_______ 8. Is Scott a teacher?

1) All of Sue's friends are teachers.
2) Scott is not Sue's friend.

Remember: To answer these questions think:
12TEN - ABCDE
1 alone **A**; **2** alone **B**; **T** ogether **C**; **E** ither **D**; **N** either **E**
12TEN - ABCDE

HELP ON OBTAINING SUFFICIENT INFORMATION

Remember to answer: **1-2-T-E-N/A-B-C-D-E**, that is, if the question can be answered with the information from statement **1 alone, and not 2 alone, answer A. If 2 alone is enough but not 1 alone, answer B. If 1 and 2 together are necessary to answer the question, then C is it, whereas if either alone will do the job, D's your response. . . but if nothing works then choose E.** [In general, the directions are more confusing than the questions]. But do not despair. Here is a metaphor that can be of great service. (What's a meta for?)

Imagine that you are the CEO and president of a big corporation, and some important problems have been put before you. Knowing more about your business than anyone else, you will first think of what you would need to solve each problem, and then call in your 2 vice presidents, your right and left-handed people, to see if they have what it takes to solve each problem. [You are the brains behind the operation, not the brawn].

Now, if the 1st VP (statement 1) is capable of solving the problem alone, then you can put an AD in the company paper, praising VP 1. That is, the answer could be A or D. [This we will call the High Road.]

If VP 1 can solve it alone but VP 2 can't solve the problem alone, then give the first VP an Airplane (A's the answer). If each can do it then you will need to make a Deal, (the answer is D).

If VP 1 can't solve it alone then you'd Better Call your Editor. That is, the answer will be either B, C, or E. [This we will call the Low Road] If VP 1 can't but 2 can, then VP 2 gets a Boat (answer B). If both VP's can't solve it alone but they can solve it together, then you will need some Corroboration (answer C). At last, if neither can solve it then you will be left Empty handed (answer E).

Remember: you don't have to solve the problem. You just need to know that it **can** or **cannot** be solved with the additional information. If the question involves finding an exact value then you will generally need as many equations as there are variables in the problem. If the question asks for a relation or ratio then think what would be required for you to go to the United Kingdom and get the problem solved.

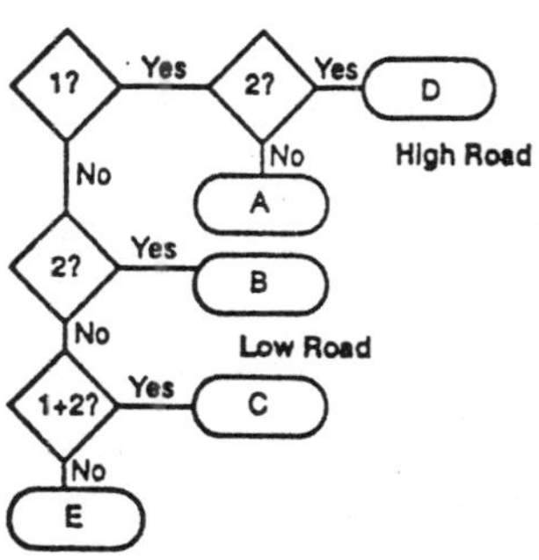

√ If you have a quadrilateral (a 4 sided polygon) then one pair of equal parallel sides, or 2 pairs of parallel sides is sufficient to declare it a parallelogram. IF you have a parallelogram, one right angle makes it a rectangle. IF you have a rhombus, **one** right angle makes it a square.

√ Confucius say: *Be careful of VP who brings up extraneous information.*: If the problem is about per cents you will need 2 out of three: is, of and %. If it's about circles remember if you know one: diameter, radius, circumference or area . . . you know all.

Enjoy your position -- have fun with these problems. It is good practice for your future career.

Example If p and q are integers, is p less than q?

1. The cube of p is less than the cube of q.
2. The square of q is greater than the square of p.

(A) Statement 1 alone is sufficient to answer this question, but 2 alone is not sufficient.
(B) Statement 2 alone is sufficient to answer this question, but 1 alone is not sufficient.
(C) Both statements together are sufficient to answer the question but neither alone is sufficient.
(D) Either statement 1 or 2 alone is sufficient to answer the question.
(E) The is not enough information to answer the question.

Answer If p and q were positive then we could answer this question with either 1 or 2. However, if p and q were negative, say p = -3 and q = -2, then cubing both would yield the correct relationship since -27 < - 8 and p < q, but squaring both would give us the reverse relationship since 9 > 4. Thus, since the cube of a number retains the same sign as the original number, 1 alone is sufficient to answer the question and the answer is A.

Example If a steel beam five meters long is cut into three pieces, what is the length of each piece?

1. The longest piece is the same length as the other two put together.
2. The shortest piece is 1 meter.

(A) Statement 1 alone is sufficient to answer this question, but 2 alone is not sufficient.
(B) Statement 2 alone is sufficient to answer this question, but 1 alone is not sufficient.
(C) Both statements together are sufficient to answer the question but neither alone is sufficient.
(D) Either statement 1 or 2 alone is sufficient to answer the question.
(E) There is not enough information to answer the question.

Answer There are three variables in this problem: longest, middle, and smallest piece. In general, we need three equations to solve this. We are told initially that L + M+ S = 5. Statement 1 gives us the relationship: L = S + M. Statement 2 give us the relationship that S = 1. Thus, we need both statements to answer the question, and the answer is C.

Example A circle is inscribed in a square. What is the area of the region that is inside the square but not inside the circle?

1. The circumference of the circle is 8 π.
2. The perimeter of the square is 32 square centimeters.

(A) Statement 1 alone is sufficient to answer this question, but 2 alone is not sufficient.
(B) Statement 2 alone is sufficient to answer thisquestion, but 1 alone is not sufficient.
(C) Both statements together are sufficient to answer the question but neither alone is sufficient.
(D) Either statement 1 or 2 alone is sufficient to answer the question.
(E) The is not enough information to answer the question.

Answer We need to be able to find the area of the circle and the area of the square since the difference will answer the question. If you know the circumference of a circle, you know the radius, diameter, and area, The diameter of an inscribed circle equals the length of a side of the square. Thus, we can solve this with 1 alone. Also the perimeter of a square gives us the side of a square which is also equal to the diameter, so we can solve this with 2 alone. The answer is D.

QUANTITATIVE COMPARISONS

In this section you are asked to compare a quantity in column A to a quantity in column B. Having learned the fourteen major concepts, you are in an excellent position to do extremely well in this section. Many times you will not have to work out a problem to its final solution, that is, you will be able to tell which quantity is larger without arriving at an exact answer.

Some additional information might prove valuable to make this section even easier.

Algebra

A. Quantity A is larger than B ($A > B$) if there is a positive number that can be added to B to make A. For example, $7 > 4$ since $4 + 3 = 7$. Thus if $x + 3 = y$, then y would have to be greater than x.

B. Negative exponents indicate fractions. $3^{-1} = 1/3$, $3^{-2} = 1/9$. In general, $x^{-m} = 1/(x^m)$. (Remember: Some people are negative about fractions.)

C. If they don't give you information about a variable, then it might be a fraction, negative, positive or 0. Test these four conditions before arriving at a generalization.

Geometry

A. In a triangle, the bigger angle lies opposite the bigger side.

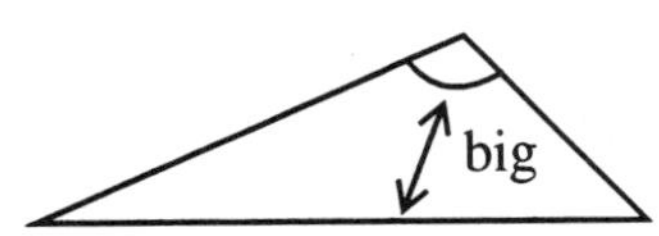

This is why the hypotenuse of a right triangle is always the biggest side of a right triangle. (Remember: hips (hyp) are wider than legs.)

B. The shortest distance between two points is a straight line.

C. The longest side of a triangle is shorter than the sum of the two other sides.

TRY THESE

Use the following statements to answer questions 1 to 4.

A) quantity in column A is greater
B) quantity in column B is greater
C) two quantities are equal
D) relationship cannot be determined from the given information

	COLUMN A	COLUMN B
___ 1.	The average of 22, 26, 28, 30	The average of 17, 27, 24, 30, 37
___ 2.	$3 + 24(3 - 2)$	$27 + 5(0)(4)$
___ 3.	3^{-3}	$1/3$

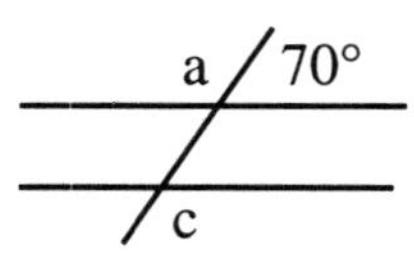

	COLUMN A	COLUMN B
___ 4.	c	a
___ 5.	1^{16}	$(-1)^{32}$

A fair coin is tossed two times

	COLUMN A	COLUMN B
___ 6.	The chance of getting 2 heads	The chance of getting no heads
___ 7.	2×3^{10}	3×2^{10}

$y/x = 1/5$

	COLUMN A	COLUMN B
___ 8.	x	5y
___ 9.	$\dfrac{1}{\frac{10}{7} - 1}$	2

DATA INTERPRETATION

In this section, you will be asked to gather information from tables or graphs and then apply the fourteen concepts to determine the answers. Ratio and Proportions will be your most used concept in this section. Chances are you will be called upon to find the percent increase or percent decrease over a period of time. If in 1950 someone made 500 dollars a month and in 1970 they made 2,000 dollars a month, then to find the percent increase we form the following proportion:

$$\frac{\text{Actual increase}}{?} = \frac{\text{starting amount}}{100}$$

$$\frac{1500}{?} = \frac{500}{100}$$

$? = 300$, that is a 300% increase

On a line graph, the greatest percent increase is where the line is the steepest and points above the horizontal axis.

The smallest percent increase would be a line that is closest to being parallel to the horizontal axis but still points above the axis.

A large percent decrease is indicated by a steep line that points below the horizontal axis.

You may wish to divide the units into smaller parts than given. Cutting each unit in half would help you get a better approximation. Using your pencil as a straight line to read the values is also very helpful.

TRY THESE

Use the chart on the next page.

1. What is the percent increase in population from1900 to 1960? (80 to 180)

   a) 100 b) 110 c) 125 d) 120 e) 135

2. In how many years after 1900 did the population double?

   a) 20 b) 30 c) 40 d) 50 e) 60

3. The largest increase in population was in the decade ending:

   a) 1960 b) 1950 c) 1940 d) 1930 e) 1920

4. What is the approximate ratio of the population in 1960 to 1930?

a) 8 : 5 b) 6 : 5 c) 3 : 2 d) 6 : 3 e) 10 : 7

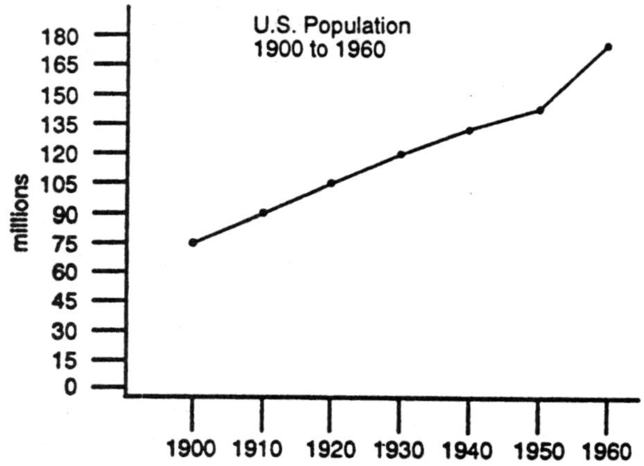

WHO OWNS THE ZEBRA?

Test Your Logic on This Brain Buster!

In New York, stranger accosts stranger with a mimeographed sheet of paper and the question "Have you seen this?" In university dormitories, the problem is tacked to doors. In suburban households, the ring of the telephone is likely to herald a voice that asks, "Is it the Norwegian?" The cause of the excitement is the brain teaser below. The facts essential to solving the problem–which can indeed be solved in combining deduction, analysis and sheer persistence–are as follows:

1. There are five houses in a row, each of a different color and inhabited by people of different nationalities with different pets, drinks and vegetables.
2. The English person lives in the red house.
3. The Spaniard owns the dog.
4. Coffee is drunk in the green house.
5. The Ukrainian drinks tea.
6. The green house is immediately to the right (your right) of the ivory house.
7. The lettuce eater owns snails.
8. Cabbage is eaten in the yellow house.
9. Milk is drunk in the middle house.
10. The Norwegian lives in the first house on the left.
11. The person who eats okra lives in the house next to the person with the fox.
12. Cabbage is eaten in the house next to the house where the horse is kept.
13. The spinach eater drinks orange juice.
14. The Japanese person eats parsley.
15. The Norwegian lives next to the **blue** house.

Now, who drinks water? And who owns the zebra? You'll find the answer in **Math Advancement**, but don't weaken until you've worked out an answer on your own!

A Solution to: Who Owns the Zebra?

FIRST	SECOND	THIRD	FOURTH	FIFTH

Yellow (2)	Blue (1)	Red (4)	Ivory (4)	Green (4?)
Norw (1)	Ukrain. (5?)	English. (4)	Spaniard (5)	Japanese (5)
Fox (5)	Horse (3)	Snails (5)	Dog (5)	Zebra (6)
Water (6)	Tea (5)	Milk (1)	O.J. (5)	Coffee (4)
Cabbage (3)	Okra (5)	Lettuce (5)	Spinach (5)	Parsley (5)

(1) — Directly Given
(2) — Deduced
(3) — Deduced
(4?)— Guess
(4) — Deduced from Guess
(5?)— Guess
(5) — Deduced from Guess
(6) — Deduced

Who Stole the Cookie from the Cookie Jar?

There are 6 cookies in 3 bags. One bag contains two chocolates, one bag contains 2 vanillas, and one bag contains one chocolate and one vanilla. All the bags were labeled. Every bag was mislabeled.

Design an experiment in which you will take ONE cookie out of ONE bag, look at it, put it back in the SAME bag, and then correctly label each bag.

CC	CV	VV
1	2	3

ANALYTICAL REASONING AND LOGICAL PUZZLES (GRE® TEST)

The typical logical puzzle includes 4 - 6 questions based on a passage or set of sentences. The puzzles vary in length, style, and difficulty. It would be a good idea to read and take notes at the same time, and try to get to the questions as quickly as possible. We will now look at the most typical kinds of logical puzzles and some effective methods for diagramming each one.

THE ER-PUZZLES

This type of puzzle involves ord**er**. You will be called upon to find some type of an arrangement of objects or people.

KEY WORDS TO LOOK FOR: BIGGER, BETTER, FASTER, CREATIVER

Sometimes: There is only one arrangement that will work and you will be able to find that from the **o**riginal **i**nformation (**O.I.**)
SOMETIMES: The **O.I.** offers you know discernible order but gives you only relationships such as A must follow B; A is not third; or B is smaller than C. For this type of puzzle, you will find the arrangements that you need from the Questions, (Q's). **EACH TIME YOU MAKE AN ARRANGEMENT FOR A Q BE CERTAIN THAT THE O.I. IS NOT VIOLATED.**
SOMETIMES: There are several possible arrangements and they will change from **Q to Q**. It is quite possible that the arrangement in question 12 is different than the arrangement in question 10.

ARRANGEMENTS CAN CHANGE PER QUESTION BUT the O.I. WILL USUALLY REMAIN THE SAME. If THE O.I. CHANGES IN A Q THEN CHANGE IT BACK FOR THE NEXT Q, UNLESS TOLD THAT EXACT INFORMATION AGAIN. FOR EACH ARRANGEMENT CHECK TO SEE THAT O.I. IS SATISFIED. THERE IS A DIFFERENCE BETWEEN THE SMALLEST AND SMALLER THAN ALL THE REST. What is it?

There are two types of diagrams that can aid you in solving these puzzles:

FLEXIBLE GRAPHS

These are used when no exact positions are indicated, only relationships such as Jane is taller than Sam. Note that Jane cannot be the smallest nor can Sam be the tallest, but Jane could be the tallest and Sam could be the shortest.

LOCKED IN GRAPHS

This is used when you know exact positions.
The sentence below tells us that A and B are adjacent to each other.
Ann IS TALLER THAN Bob AND NO ONE IS BOTH TALLER THAN B AND SHORTER THAN A.

**If you are told exact relationships such as Jan is 5 years younger than Dan, then choose a number for one of these characters, say make Dan 20 yrs old, then you can determine that Jan must be 15 yrs old.

Attribute or Matching Puzzle

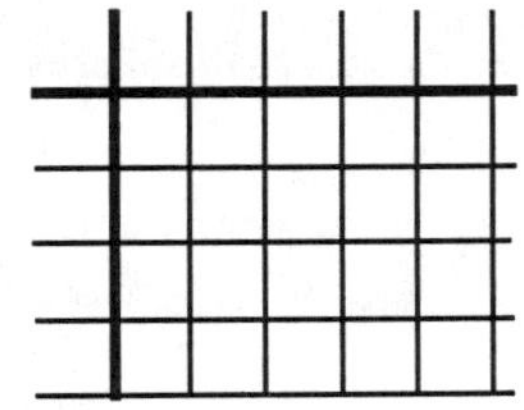

This type of puzzle is trying to match people or objects with other people, objects, professions, etc. It could be asking you to match people with houses, people with occupations, couples with dining rooms, etc.

A table composed of rows and columns could help you solve these puzzles. Let the columns be fixed positions - homes, occupations, etc. Fill in all the information you know first, and all negative information next.

TIPS FOR ANALYTICAL REASONING

For All:

- Abbreviate
- Definites first
- Relatives next
- Negatives last
- Use margins
- Only you need to read
- When using process of elimination, list orderly
- Be flexible

Types of Logical Puzzles

Venn Diagram

- All A's are B's
- Some C's are D's
- No F's are G's

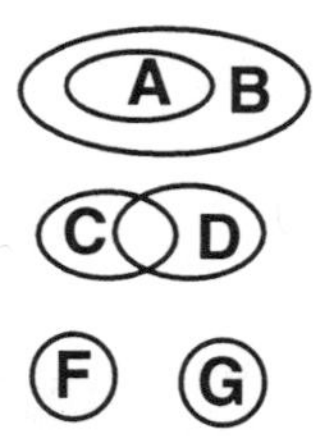

Flow Chart

- If A then B
  = A → B
- To C . . . you must D . . .
  = D → C
- Sequence of Actions

Family Trees

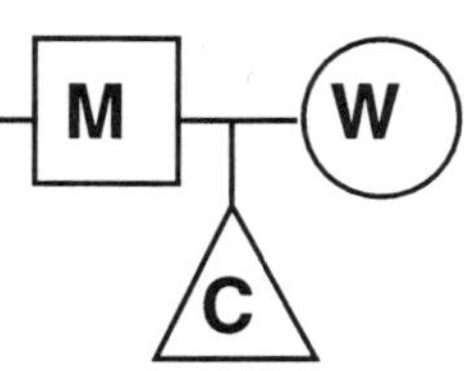

Linear

- Given – diag/symbols
  (What will not change)
- Definites
- Relatives
- "Not's"
- Process of elimination
- Assign Numbers

Groupings

- Given – diag/chart/symbols
- AB or AC = A works only with B or C
- M ≠ N means M does not equal N
- A ≠ 1, A can't go first
- B ← C, C follows B always
- If ≠ No one can follow themselves
- Diag change/situation

TO ANSWER QUESTIONS QUICKLY:				
		ALWAYS	SOMETIMES	NEVER
1.	I read the initial information once and translate each fact into symbolic form while reading.			
2.	I go directly to the questions after reading and translating the initial information.			
3.	I talk to myself only about the question.			
4.	I relate back to the symbols for information to solve the problems.			
5.	When I find a correct answer, I go immediately to the next question.			
6.	I give myself only positive feedback.			

TO ANSWER QUESTIONS CORRECTLY:				
		ALWAYS	SOMETIMES	NEVER
1.	I check to see the initial conditions are met.			
2.	I use only the initial information and the information in the question to answer that question.			
3.	I cross out conditions that could not exist.			
4.	I use the process of elimination when necessary.			
5.	I underline words like: cannot, must, could, possible, never, can, is, may.			

IF P THEN Q: $P \rightarrow Q$

Mathematics relies heavily on a system of logical reasoning called deduction. Deduction is a method of proof based on reasoning rather than on experiences, measurements, or trial and error. Deduction results from thinking rather than from experimenting.We reason logically by using a set of statements. These may be simple statements like "Today is Friday," or they may be compound such as "Today is Friday and it is raining outside." To form a compound sentence, we use connecting words and symbols such as: and ($\wedge$); or ($\vee$). To take the opposite of a sentence we use words and symbols like: not ($\sim$).

In mathematics we often use letters to represent numbers. Likewise, in logic we use letters and symbols to represent statements. The expression $P \rightarrow Q$ is a compound statement called an implication or conditional statement. If P occurs, it is implied that Q will also occur. To demonstrate some important notations and rules in logic let us begin with:

P = Andy lives in Chicago.

Q = Andy lives in Illinois.

We can then take the opposite of these sentences by using the symbol: $\sim$

$\sim$ P = Andy does not live in Chicago.

$\sim$ Q = Andy does not live in Illinois.

Let us start with the compound conditional statement $P \rightarrow Q$:

$P \rightarrow Q$ = If Andy lives in Chicago, then Andy lives in Illinois. No matter who Andy is, we know that this statement is TRUE, i.e. if Andy were to live in Chicago he would also have to live in Illinois.

$Q \rightarrow P$ = If Andy lives in Illinois then Andy lives in Chicago is the CONVERSE of the original statement. Notice that this statement is FALSE. It is quite possible for Andy to live in Illinois, say Springfield, and not live in Chicago.

$\sim P \rightarrow \sim Q$ = If Andy doesn't live in Chicago, then Andy doesn't live in Illinois is the INVERSE of the original statement. Notice that this, too, is FALSE. Again, Andy could live in Highland Park which is another city in Illinois.

$\sim Q \rightarrow \sim P$ = If Andy doesn't live in Illinois, then Andy doesn't live in Chicago is called the CONTRAPOSITIVE of the original statement. This statement is TRUE. If Andy doesn't live in Illinois, then he cannot live in Chicago.

In logic we can always reason from the contrapositive. This is referred to the *law of contraposition.* This means that if the original statement, $P \rightarrow Q$, is true, the contrapositive, $\sim Q \rightarrow \sim P$, will also be true. This can be very useful when solving logical puzzles where we have conditional statements. For example, if we are given the statement: If Mary is not on the red team then Sue will be on the green team. $\sim M(rt) \rightarrow S(gt)$.

This allows us to conclude that if Sue turns up on the yellow team, then Mary will have to be on the red team: That is, the contrapositive $\sim S\ (gt) \rightarrow M\ (rt)$ must also be true. This is like getting "double" your value from conditional statements!

Puzzle One

In downtown Olympia there are six large buildings. They are called: Alpha, Beta, Chi, Delta, Epsilon, and Fi. From shortest to tallest, they have been painted: yellow, white, green, purple, rose, and orange, respectively.

Alpha, which is not rose, is taller than Chi.
Beta is taller than Epsilon and Fi, but shorter than Delta.
Chi is taller than Fi.

1. If Chi is shorter than Epsilon, what color must Fi be?

   a) Yellow b) white c) green d) purple e) rose

2. Which of the following is a possible arrangement of the buildings from shortest to tallest?

   a) Epsilon, Chi, Fi, Alpha, Beta, Delta
   b) Fi, Chi, Alpha, Epsilon, Beta, Delta
   c) Fi, Chi, Alpha, Beta, Epsilon, Delta
   d) Fi, Chi, Epsilon, Beta, Alpha, Delta
   e) Fi, Epsilon, Chi, Beta, Alpha, Delta

3. If Alpha is Green, Chi must be:

   a) Yellow b) white c) green d) purple e) rose

4. Which of the following cannot be a possible arrangement of buildings from shortest to tallest?

   a) Epsilon, Fi, Chi, Alpha, Beta, Delta
   b) Epsilon, Fi, Chi, Beta, Delta, Alpha
   c) Fi, Epsilon, Beta, Delta, Chi, Alpha
   d) Fi, Epsilon, Chi, Alpha, Beta, Delta
   e) Fi, Epsilon, Delta, Chi, Beta, Alpha

5. If Epsilon is purple, it is possible for Beta to be which of the following colors:

   I. Green II. Rose III. Orange

   a) I only b) II only c) I and II only d) II and III only e) I, II, and III

6. If Delta is rose, it is possible for Beta to be which of the following colors?

I. White II. Green III. Purple

a) I only b) II only c) III only d) I and II only e) II and III only

Puzzle Two

Ann, Bea, Carl, Dan, Earl and Fay were all born on April 1st at noon.

Ann is five years older than Earl
Bea and Dan are four years apart in age
Earl is three years younger than Dan
Carl is two years older than Ann
Dan and Fay are three years apart in age

1. It is possible for which of the following persons to be the same age?

I. Ann and Bea II. Bea and Carl III. Bea and Earl

a) I only b) II only c) III only d) I and II only e) II and III only

2. Which of the following must be true?

I. Ann and Fay are five years apart II. Dan and Ann are two years apart
III. Earl and Fay are six years apart

a) I only b) II only c) III only d) I and II only e) I and III only

3. If Bea is an even number of years old, which of the following must be true?

a) Earl is an even number of years old?
b) Fay is an even number of years old?
c) Ann is an odd number of years old?
d) Dan is an odd number of years old?
e) Fay is an odd number of years old?

4. If no two of these six persons are the same age, which of the following is the correct order of persons from youngest to oldest?

a) Bea, Earl, Dan, Ann, Fay, Carl
b) Bea, Fay, Earl, Dan, Ann, Carl
c) Earl, Dan, Ann, Fay, Bea, Carl
d) Earl, Fay, Dan, Ann, Carl, Bea
e) Fay, Earl, Dan, Ann, Bea, Carl

5. If at least three people in the group are older than Fay, which of the following must be true:

I. Carl and Bea are the same age. II. Bea and Fay are the same age.
III. Fay and Earl are the same age.

a) I only b) II only c) III only d) I and II only e) II and III only

6. Which of the following person could be younger than all the others?

I. Bea II. Earl III. Fay

a) I only b) II only c) III only d) I and II only e) II and III only

Puzzle Three

On a particular evening, four men: Mike, Ned, Otis, and Paul and five woman: Ray, Sue, Tina, Ursula, and Vanessa, will dine at the same restaurant. There are only five private dining rooms available that can accommodate at most one couple. Each of the men will dine with one of the women in one of the private dining rooms.

The dining rooms are numbered 1, 2, 3, 4, and 5.
Ned will not dine with Ray and Mike will not dine with Ursula
Otis must dine with Tina or Vanessa
Mike will dine in room 5
Ned will dine in room 2, and Vanessa will dine in room 1.

1. If Tina dines in room 5 who might Otis dine with?

a) Ray b) Sue c) Tina d) Ursula e) Vanessa

2. If Paul dines in room 3, and Otis cannot dine in room 1, then who is in room 4?

a) Sue b) Ray c) Vanessa d) Tina e) Ursula

3. If Otis dines with Vanessa, where may Paul dine?

I. Room 1II. Room 4 III. Room 3

a) I only b) II only c) III only d) I or II only e) II or III only

4. If Sue dines in room 2, and Otis dines with someone in room 4, who must dine with Mike?

a) Ray b) Sue c) Tina d) Ursula e) Vanessa

5. If Ned dines with Sue, with whom may Mike dine with?

I. Ray II. Tina III. Ursula IV. Vanessa

a) II only b) I or II only c) I or IV only d) I, II, or IV only e) I, II, III, or IV

6. If Tina dines in room 2, who may dine in room 5?

I. Ray II. Sue III. Ursula IV. Vanessa

a) I only b) IV only c) I or II only d) II, or IV only e) I, II, III, or IV

Puzzle Four

Five cities - Central, Dannon, Eden, Fordham, Ganon - are ranked from first (best) to fifth (worst) on each of three criteria: culture, finances, lifestyle.

For each of these three criteria, none of the five cities received the same ranking as any other city.
In culture, Ganon ranked third, and Fordham fourth.
In finances, Dannon is ranked second, Eden third, Central fourth.
In lifestyle, Ganon's rank is better than Central's, Dannon is ranked fourth, and Fordham fifth.
Eden receives a better ranking in culture than in lifestyle.
Dannon's three rankings are all different from each other.

1. Which one of the following **is** a complete and accurate list of the rankings any one of which could be the ranking on culture given to Eden?
   a) first b) first, second c) first, fifth d) second, fifth e) first, second, fifth

2. Which one of the following is a town that **cannot** be ranked fifth on any one of the three criteria?
   a) Central b) Dannon c) Eden d) Fordham e) Ganon

3. Which of the following **could be** true?
   a) Central is ranked first in both culture and lifestyle.
   b) Dannon is ranked first in both culture and finances.
   c) Eden is ranked first in culture and third in both finances and lifestyle.
   d) Fordham is ranked fifth in lifestyle and fourth in both culture and finances.
   e) Ganon is ranked third in both culture and lifestyle.

4. If Dannon is ranked first in culture, then it **must be** true that:
   a) Central is ranked second in culture.
   b) Central is ranked third in lifestyle.
   c) Eden is ranked second in lifestyle.
   d) Eden is ranked third in lifestyle.
   e) Ganon is ranked fifth in finances.

5. If Central is ranked second in culture, then which of the following **can be** true?
   a) Central is ranked second in lifestyle.
   b) Dannon is ranked first in culture.
   c) Eden is ranked first in lifestyle.
   d) Eden is ranked fifth in culture.
   e) Ganon is ranked third in lifestyle.

6. If Ganon is ranked first in finances and Eden is ranked second in lifestyle, then it **is possible** to deduce with certainty all three rankings for exactly how many of the towns?
   a) one b) two c) three d) four e) five

7. Which of the cities **cannot have** the rank for the area given.
   a) Central - 1st Culture
   b) Dannon - 5th culture
   c) Eden - 3rd lifestyle
   d) Fordham -1st finances
   e) Ganon - 2nd lifestyle

Puzzle Five

Six students, Juan, Karl, Laura, Meg, Oliver, and Paul, participate in a track meet. Two of them will enter the marathon, two enter the javelin, and two enter the relay.

Each participant enters only one event.

If Karl enters the marathon, then both Juan and Laura enter the relay, and if both Juan and Laura enter the relay, Karl enters the marathon.

If Meg doesn't enter the javelin, then Paul will enter the relay.

If Oliver enters the relay, then Juan does not enter the relay.

If Oliver enters the relay, then Paul does not enter the relay.

1. If Karl enters the marathon, then which statement cannot be true?
   a) Juan enters the relay and Meg enters the javelin.
   b) Oliver enters the marathon and Meg enters the javelin.
   c) Paul enters the marathon and Oliver enters the javelin.
   d) Paul enters the relay and Oliver enters the javelin.
   e) Paul enters the marathon and Laura enters the relay.

2. If Meg enters the marathon, which statement cannot be true?
   a) Both Karl and Oliver enter the Javelin.
   b) Karl enters the relay and Oliver enters the javelin.
   c) Karl enters the marathon and Laura enters the relay.
   d) Juan enters the marathon and Laura enters the javelin.
   e) Juan enters the marathon and Karl enters the relay.

3. If both Karl and Oliver enter the relay, which statement must be true?
   a) Juan and Paul enter the marathon.
   b) Laura enters the marathon.
   c) Laura enters the javelin.
   d) Paul enters the javelin.
   e) Meg enters the javelin.

4. If both Juan and Karl enter the same event then which statement can be true?

   I. Juan enters the relay II. Laura enters the marathon III. Oliver enters the relay

   a) I only b) II only c) III only d) I and II only e) I, II, and III

5. If Meg and Oliver each enter different events which statement cannot be true?
   a) Karl and Oliver enter the marathon.
   b) Juan and Meg enter the relay.
   c) Laura and Oliver enter the relay.
   d) Meg and Paul enter the relay.
   e) Oliver enters the javelin and Paul enters the relay.

6. If both Juan and Karl enter the relay, which statement must be false?
   a) Laura and Meg enter the marathon
   b) Laura and Oliver enter the marathon.
   c) Meg and Paul enter the javelin
   d) Meg and Laura enter the javelin.
   e) Laura enters the javelin and Paul enters the marathon.

MORE TIPS FOR LOGICAL PUZZLES

Remember

Underline quantifying terms such as: Never, could, possible, must, can, can't...

- To answer questions involving situations that could **Never** (**Can't**) happen or situations that are **Not** possible, look for answers that contradict the original information, then check to see that each spot that must be filled is filled, and finally that the arithmetic works.
- To answer questions that **Could** (**Can**, **May**) Be True, or **is/ are Possible**, after eliminating ones that violate the original information, check the remaining answers in order. If A is a scenario that CAN work you have found the answer! Do not check further!
- To answer questions that **MUST** be true requires a bit more work. Once you eliminate scenarios that violate the initial conditions you must check that no other combination or permutation is possible. 1.e. suppose it asks you if the Banker must be second. Once you know the Banker **COULD** be second, you need to be sure that there is no other place for that banker to be.

Remember

Do not use information obtained in one question to answer another question, unless you have the exact same information.
Everything from the original information must work for every question. Everything has to fit!
Sometimes the easier questions come last.
BE FLEXIBLE- sometimes the puzzles are combinations of types.
YOU CAN DO IT!!!! THINK POSITIVELY!!!! ENJOY!!!

After You Complete A Puzzle Answer the Following:

1. I now know:

   (a) To look for:

   ____

   ____

   (b) Not to do the following:

   ____

   ____

   (c) The meaning of:

   ____

   ____

2. What will you do the same when you see a similar puzzle to this one?

____

____

3. What will you do differently when you see a similar puzzle to this one?

____

____

4. How do you feel about this type of puzzle?

____

____

5. If you were to write a puzzle that was similar to this one, what would it include?

____

____

PUZZLE SIX

Larry, Mark, Nick, Oliver, and Paul are wrestlers who might participate in a tournament in Germany.

> If Larry is invited he will attend.
> If Mark is invited he will attend only if Paul will also attend
> If Nick is invited he will attend only if Oliver attends.
> If Oliver is invited he will attend only if Larry does not attend
> If Paul is invited he will attend only if Nick does not attend.

1. If Larry, Mark, Nick, Oliver, and Paul are all invited to Germany, how many will actually attend?

   (A) one (B) two (C)three (D) four (E) five

2. If Only Mark, Nick, Oliver and Paul are invited to Germany , how many of them will attend?

   (A) zero (B) one (C) two (D) three (E) four

3. If only Larry, Nick, Oliver and Paul are invited to Germany, how many of them will attend?

   (A) zero (B) one (C) two (D) three (E) four

4. If only Mark, Nick, and Paul are invited to Germany, which of them will attend?

   (A) only Mark (B) Only Nick (C) only Paul (D) Mark and Paul,
   (E) Mark, Nick, and Paul

5. If only Larry, Mark, and Nick are invited to Germany, which of them will attend?

(A) only Larry (b) only Mark (C) only Nick (D) Larry and Mark
(E) Larry and Nick

6. Which of the following wrestlers could not attend together?

(A) Larry and Mark (B) Larry and Nick (C) Larry and Paul (D) Mark and Paul
(E) Nick and Oliver

7. If the sponsors wanted Mark and Oliver to attend who would have to be invited?

(A) Larry, Matt, and Oliver only (B) Mark, Nick, and Oliver only
(C) Mark, Nick and Paul only (D) Mark, Oliver, and Paul only
(E) Mark and Oliver only

PUZZLE SEVEN

Five males, Carl, Rob, Mark, Juan, and Larry and two females Deb and Toni each took private courses in geometry and/or algebra with Dr. Simon over a five year period from 1980 through 1984. These were the only private students that Dr. Simon had during this time, and she had only two students each year- one in Algebra and one in Geometry.

No person took the same course more than one time
No person took Algebra and Geometry together
No two people took the same course at the same time.
No person took Algebra after they took Geometry.
The males did not take Geometry from Dr. Simon in consecutive years.

1. How many people took both Algebra and Geometry from Dr. Simon?

(A) zero (B) one (C) two (D)three (E) four

2. The person who took Algebra in which year could not have taken Geometry from her?

(A) 1984 (B)1983 (C)1982 (D)1981 (E) 1980

3. If Rob took Algebra before Deb did, it is possible to determine the identify of the person who took Geometry in which year(s)?

(A) 1981 and 1983 (B) 1980 and 1983 (C) 1983 only (D) 1981 only (E) 1980 only

4. If there were at least two years between the when Carl finished Geometry and Larry began taking Geometry which of the following is a complete list of the people who could have taken Algebra in 1984?

(A) Mark and Juan (B) Mark, Juan and Rob (C) Mark, Juan, Rob and Carl
(D) Mark, Juan, Rob, Carl, and Larry (E) Mark, Juan, Rob, Deb, and Toni

5. If there were at least two years between the when Carl finished Geometry and Larry began taking Geometry, and if Mark did not take Algebra from Dr. Simon, which of the following is a complete and accurate list of those who took Geometry with Dr. Simon.

(A) Carl, Larry, Mark, Deb, Rob
(B) Carl, Larry, Mark, Deb, Toni
(C) Carl, Larry, Mark, Deb, Juan
(D) Carl, Larry, Mark, Toni, Rob
(E) Carl, Larry, Mark, Toni, Juan

6. If there were at least two years between the when Carl finished Geometry and Larry began taking Geometry, and if Mark did not take Algebra from Dr. Simon, then who must have taken Algebra from Dr. Simon

(A) Larry and Deb only
(B) Deb and Toni only
(C) Larry, Deb, and Toni only
(D) Larry, Deb, Toni, and Rob only
(E) Larry, Deb, Toni, Rob, and Juan

PUZZLE EIGHT

Mary is going on a ski trip. She will need exactly three sweaters for her trip. The sweaters come in three colors: red, white and blue, and with three logos on the pocket: tree, flower, and bird.

A sweater is identified by its color and logo
Mary will not buy two identical sweaters
Mary will not buy both a red sweater and a blue sweater
No red trees are available
No blue birds are available.

1. Which of the following must be false?
(A) two of the sweaters that Mary will buy are red and two are trees
(B) two of the sweaters are white and two are trees
(C) two of the sweaters are blue and two are trees.
(D) two of the sweaters are red, one is flower and one is bird
(E) two of the sweaters are white, one is flower, one is bird

2. If Mary buys a red bird shirt, which must be false?
(A) She buys two flower sweaters
(B) She buys two tree sweaters
(C) She buys two flower sweaters
(D) She buys two red sweaters
(E) She buys two bird sweaters

3. If Mary does not buy a white flower sweater, which sweater(s) must she buy?
(A) white tree sweater or a red bird
(B) white tree or white bird
(C) blue tree or red bird
(D) blue tree or white tree
(E) blue flower or red flower

4. If Mary buys exactly one white sweater and does not buy two with the same design, then she cannot buy which of the following?
(A) white tree
(B) white flower
(C) white bird
(D) blue tree
(E) blue flower

5. If neither blue tree nor red bird are available, which must Mary buy?
(A) small sweater
(B) white flower
(C) either a blue or a red
(D) either a white tree or a white bird
(E) either a blue flower or a white bird

PUZZLE NINE

Brandy is testing out model airplanes to exhibit at a convention. The airplanes are labeled A through F.

Airplane A is faster than Airplane B, and B is faster than C
Airplane A is faster than D, and D is faster than E
Airplane F is not faster than C

1. How many airplanes must be slower than A?

(A) five (B) four (C) three (D) one (E) zero

2. Which of the following statements could be true?
I. Exactly two planes are slower than B
II. Exactly three planes are slower than B
III. Exactly four planes are slower than B

(A) I only (B) II only (C) III only (D) II and III only (E) I, II, and III

3. Which of the following statements could be true?
I. B is faster than E
II. B and E are the same speed
III. B is slower than E

(A) I only (B) II only (C) III only (D) I and II only (E) I, II, and III

4. Which of the following statements could be true?
I. The difference in speed between B and C is less than the difference in speed between B and F.
II. The difference in speed between B and C equals the difference in speed between B and F.
III. The difference in speed between B and C is greater than the difference in speed between B and F.

(A) I only (B) II only (C) III only (D) I and II only (E) I, II, and III

5. If a seventh airplane, G, were faster than exactly five of the six planes described above, which of the following conclusions must be true?

(A) G is faster than A
(B) G is faster than B
(C) G is slower than A
(D) G is slower than B
(E) G is slower than exactly one of the other six

6. If a seventh airplane G, were faster than exactly three of the other six planes and slower than exactly three of the other six planes, which of the following conclusions must be true?

(A) G is faster than C
(B) G is faster than F
(C) G is slower than B
(D) G is slower than D
(E) G is slower than E

PUZZLE TEN

Seven people are waiting on line at the DMV.

The first person on line has the same last name as the last person on line.

The number of people who were between the two Smiths is one more than the number who were after Abbott.

Barker was before Plum.

Two people were named Lopez.

1. If the second person on line was a Smith, which must be true
   I. The sixth person was Abbott
   II. The third person was Barker
   III. The fifth person on line was either a Smith or Abbott

(A) I only (B) II only (C) I and II only (D) II and III only (E) I, II, and III

2. Each of the following describes a possible order on line except:

(A) both Smiths were before Abbott
(B) both Lopez's were before Abbott
(C) both Lopez's were before Plum
(D) at least one Smith was before Barker
(E) Plum was before Abbott

3. If the first three people on line were female, which must be true?

(A) The last person on line was a man
(B) Only three people on line were women.
(C) At least one Smith was a women
(D) At least one Lopez was a women.
(E) Abbott was a man.

4. If Plum was immediately after Barker, and Abbot was before Barker, which of the following must be true?
   I. The second person on line was a Lopez
   II. The last person on line was a Smith
   III. The third person on line was Abbott

(A) I only (B) II only (C) III only (D) I and II only (E) I, II, and III

5. If one Lopez was in back of another Lopez, the second person on line was:

(A) Abbott (B) Barker (C) Plum (D) a Smith (E) a Lopez

6. Assume that one of the Lopez's was married and the other was single, and the single was ahead of the married one and immediately after an Abbott, how many arrangements would satisfy this and the other conditions?

(A) one (B) two (C) three (D) four (E) five

7. If the first person on line was Barker, which of the following conditions could not be true?

(A) Seven people were on line.
(B) the first person had the same last name as the last person.
(C) The number in line between the two Smiths is one more than the number in line after Abbott.
(D) Barker was before Plum.
(E) Two people were named Lopez.

MATH MASTERY

1. In the figure, M is the midpoint of the base of ▱RSTU. What is the ratio of the area of RSM to the area of the parallelogram?

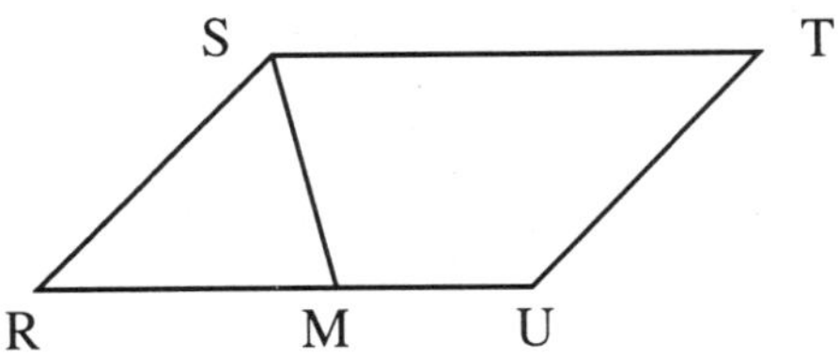

a) 1 : 2 b) 1 : 4 c) 1 : 3 d) 2 : 5 e) 2 : 7

2. The fraction u/v (u and v positive) will have a value greater than 3 if:
a) 3u – 3v b) u > v c) u > 3 d) u > 3v e) 3v > u

3. If 35rt + 8 = 42rt, then rt =

a) 8/7 b) 7/8 c) 8/87 d) 87/8 e) –8/7

4. An \$80 suit is sold for \$68. What is the rate of discount?

a) 15% b) $17\frac{11}{17}$% c) 12% d) 20% e) 24%

5. If a triangle of base 6 has the same area as a circle of radius 6, what is the altitude of the triangle?

a) 6π b) 8π c) 10π d) 12π 3) 14π

6. A pool has the shape and dimensions shown in the figure. How many square feet in its area?

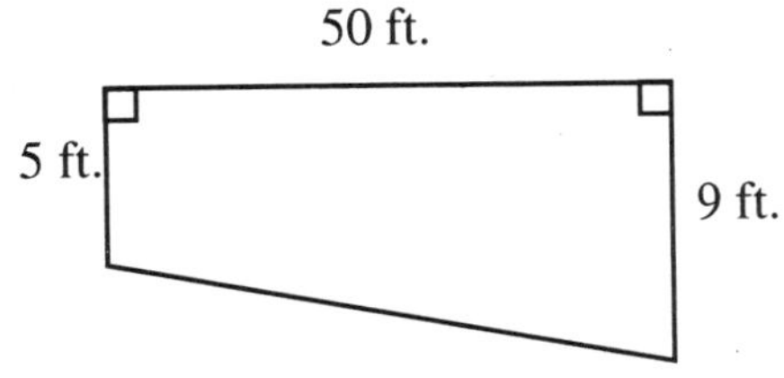

a) 300 b) 250 c) 350 d) 500 e) 530

7. If $7x = 3x + 12$, then $2x + 5 =$

a) 10 b) 11 c) 12 d) 13 e) 14

8. If PQ = QS, QR = RS and angle PRS = 100 degrees, what is the measure, in degrees, of angle QPS?

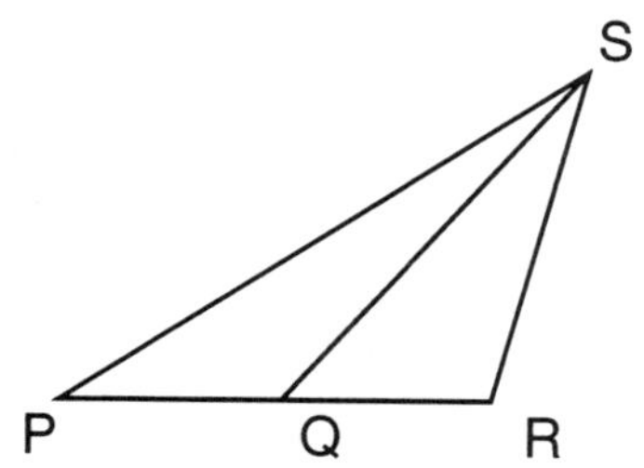

a) 10 b) 15 c) 20 d) 25 e) 30

9. A student has an average of 78% in four of his major subjects. What grade must he get in his fifth major subject, if he is to average 80 percent in all five majors?

a) 82 b) 84 c) 86 d) 88 e) 89

10. AD and BE are straight lines.

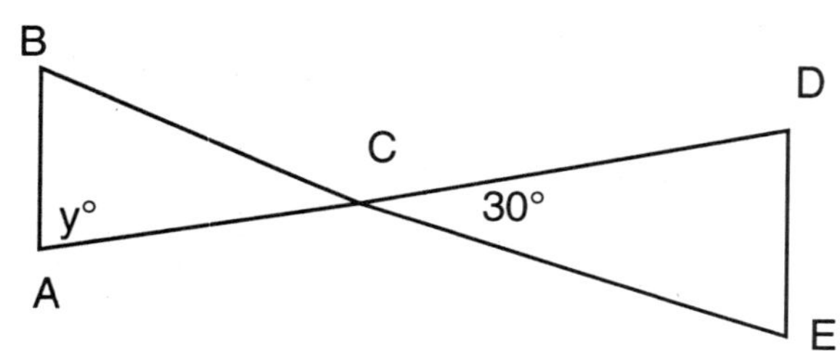

In the figure above, if BC = CA, then $y =$

a) 15 b) 60 c) 75 d) 150 e) 165

CORRECT

$\frac{\square}{10}$

PART II

MATHEMATICS TODAY AND ADVANCED MATHEMATICS

Deriving mathematics from the learner's reality is an important component in mastering mathematics. Also important is the knowledge that the power and beauty of mathematics is in its abstractness.

Math Advancement supports the theory that to learn mathematics with an understanding and appreciation of its role in the real world and to be able at the same time to see the power of its abstractness, requires using several models for the same mathematical concepts. This is illustrated in the section titled "What's Your Preference?"

Reasoning with mathematical symbols is usually easier and more efficient than working totally with physical referents.

Mathematics Today: Early Experiences with Mathematics

A Call for a Mathematically Confident and Competent Global Society

New demands are being made on the populace. There are new expectations and new employment patterns. New technology, and the subsequent rapid sociological change it produces, requires more and more mathematical sophistication, yet mathematics education remains a debacle. According to the late mathematics educator Morris Kline, in his book Why Johnny Can't Add (1977),

> One need only ask a typical college graduate how much two-thirds of three-quarters is; and if perchance s/he should give the correct answer, one need only challenge it to discover her/his uncertainty. The only generally understood fact about the subject is that it is 'ununderstandable.' Adults' reaction, based on schooling in the subject, is a mixture of awe and contempt; usually it is more of the latter . . . . Even otherwise well-educated people boast about their ignorance of the subject as if it were beneath notice. (p. 2)

It is not a new phenomenon that a large segment of the population dislikes, fears, and avoids mathematics. This situation used to be tolerable when technology was at its pre-infant and infant stages. However, today mathematics avoiders are not only cut off from full participation in society and filtered into lower paying jobs, they are also unaware of how technological progress can impact the environment, and ultimately provide solutions to some of the most challenging problems facing the global community today.

Mathematics: A Human Adventure

In our attempts to understand and, eventually overcome, the difficulties that people have experienced with mathematics, it is important to begin with a survey of what we know about the relationship between humans and mathematics.

- All humans are innately mathematical; math is a part of our genetic makeup--we are all born mathematicians. Humans will devise their own mathematics systems if left alone. They will develop art works ("primitive" art tends to be mathematical), compose music (rhythms) and establish patterns (rituals). Math learning deficiencies show up early once formal mathematics schooling begins, and all too often they lead to feelings of demoralization.
- Formalizing a bridge from the individual's instinctive mathematics to formal communications in math is where many learning problems begin. One of the things that turns children off to mathematics is the overemphasis on exactness and precision in the early stages of learning mathematics. This impedes creative explorations and establishes barriers. Other killers of the joy of mathematics include lack of estimation practice; isolation in the learning process; lack of experimentation; "drill and kill" exercises; and excessive emphasis on following arbitrary and disconnected frameworks.
- Much of present mathematics schooling is misguided and inappropriate for large segments of children; for some it is too easy. For others it is irrelevant. There has been a failure to understand where the student is "coming from"; that is, the language, culture, economic, and social conditions that affect the child's life. To reverse this trend and make mathematics accessible to all, we need to provide new educational methods and techniques and to teach mathematics that is relevant to all students.

Changing The Way Mathematics is Taught--A Note to Parents and Teachers of Mathematics

The following is a sampling of what a classroom teacher can do to secure the success of a global mathematics curriculum of benefit to all students.

- Provide a cooperative learning environment. Show all students (male and female) how to teach younger children mathematics. Every math student can become a teacher. This immediately gives status to the student doing the teaching and strengthens the ability to share for both student and "teacher."
- Explore many ways to do mathematics. A student should be able to choose between paper and pencil, calculator, computer, or mental arithmetic in all mathematical experiences, including tests.
- Enhance communication skills. Ask students to explain an algorithm, orally and in writing, by describing how they arrived at a conclusion. The process of explaining strengthens the process of learning. Have students explain a problem they know very little about to someone who knows even less.
- Have students work in groups (preferably small ones). Begin with pairs. Try to ensure minimum personality conflict. Encourage talking in groups. Discuss the nature of correcting each other because students tend to be very sensitive to each others feelings.
- Use an integrated curriculum. Present problems that require research into other disciplines and have students offer a wide range of possible solutions where mathematical modeling is called upon to explore the viability of each solution.
- Create an exploratory environment in math. Using common objects to serve as learning tools and models, students can develop creative thinking and gain greater understanding of challenges that our environment faces.
- Establish a climate where open-ended questions are frequently asked and explored. Explore news reports relating to mathematics.
- Explore many ways to understand, use, and enjoy math. Have students make mathematical connections through other disciplines and current events. Have students write about math and do math projects. Explore topics that develop from the student's curiosity. Explore the history of mathematics and the contributions of a vast array of cultures which have added to our current wealth of mathematical knowledge.

- Play mathematical games, solve logical puzzles, create new games and create new puzzles. Deliver the message: Math is FUNdamental--and it is for everyone.

Changing What is Taught in Mathematics

Changes in technology and the broadening of the areas in which mathematics can be applied have also resulted in growth and changes in the discipline of mathematics itself. Society is now in a "golden age" of mathematical production, with more than half of all mathematics having been invented since World War II. The following is a sampling of topics that need to be explored in every pre-college program.

- Understanding very big and very small numbers as a key to discovering the world. The population of the earth at the time of Jesus was perhaps 250 million people; today the number is believed to be 5.5 billion and by the turn of the century is predicted to be 7 billion. The age of the earth is well established to be 4.6 billion years. The budget deficit in the United States is over $4 trillion. Yet despite the need to use and understand numbers of this magnitude, the confusion between millions, billions, trillions (and the imaginary zillions!!) is endemic in everyday life.
- Understanding measurement. The need for a universal system of measurement is apparent from the story of a father who, after a very successful hunt, tells his son to "hide the meat so that the wolves will not get to the leftovers." When the son asks where he should hide the meat, the father says: "put it in a place higher than a person's head." The next day when the father and son come to reclaim the meat, the wolves had devoured it. The father is angry at the son for putting the meat in such a low place. (A standard system of measurement was needed since people vary in size.) As of September 30, 1993, the United States has officially adopted the metric system of measurement. It is important that all citizens understand the metric system and its impact on a global economy.
- Understanding the geometry of nature. As the world has moved on, our vision of the universe has been altered. Benoit Mandelbrot, the father of fractal geometry, declared that "Clouds are not spheres, mountains are not cones, coastlines are not circles and bark is not smooth, nor does lightning travel in a straight line." The new realization of what the geometry of nature is, may be referred to as fractal geometry . The emerging theories of chaos and *fractal geometry* (See *Chaos Theory and Fractal Geometry--Bounded Madness*) give us new insights into the environment and how to protect it. They also offer us a new lens through which to view the world and challenge, the "deterministic" model of a clockwork universe.
- Understanding and the applying new technologies. The shift in society from industrialization to data acquisition has transformed all aspects of mathematics. Computational devices such as calculators and computers make more classroom time available for discussions of mathematical substance. They can be used as tools to help students understand problematic situations through mathematical analysis. It is this author's sincere hope that all tests will allow calculator use in the near future.
- Understanding and applying theories of probability and statistics. Statistics is a method of *panning precious order from the sands of complexity.* With the proliferation of information, and its natural by-product of burgeoning data, a new responsibility has been imposed on the populace--the need to recognize erroneous and sometimes deliberately misleading arguments and to require scientists, economists, and politicians to have an almost fanatic love for the truth (see *What is the Probability Of??* and *What's Stat You SAY??*).

Mathematics is a tool, not the whole tool kit. It takes time to learn how best to use any tool, especially one of such extreme power. However, to participate fully in the world of the future, everyone must tap into the power of mathematics. Mathematics educators must take full advantage of the technology available today, and assess students' understanding of mathematics in a variety of authentic and significant ways.

In **Part II** of ***Math Advancement*** you will find some of the fundamental ways to build a strong intuitive understanding of the concepts that you will find in mathematics classes in higher education and in the world in which we inhabit.

WHAT IS THE PROBABILITY OF??

Introduction

Probability is a tool that can assist us in making the best decisions conceivable. It is a tool used by biologists, physicians, economists, social scientists, political scientists, actuaries, geologists, business administrators, and many other professionals. Since there is considerable uncertainty in decision making, it is often important that all the known risks involved be scientifically evaluated. Probability is often referred to as the Science of uncertainty." Probability allows us to measure effectively the risks in selecting one solution to a problem over alternatives.

A probability is a measure of the *likelihood* that a particular event will happen. It is a *ratio*, and thus can be expressed as a fraction, decimal, or percent (see concept 3). The likelihood that any particular event will happen may assume values between 0 (no chance of occurring) and 1 (certainty that the event will occur). A weather person can claim that there is a 70% chance of rain tomorrow, the probability of tossing a heads is $1/2$, the likelihood of Tony Gwynn getting a hit his next at bat is 0.350.

Simple Probabilities

What is the probability that a card drawn at random from a standard deck of cards will be a club? Since there are a total of 52 cards in a standard deck of cards, and 13 of these are clubs, the probability is 13/52 or 1/4. In general, the probability of an event is the number of favorable outcomes divided by the total number of possible outcomes. (This assumes the outcomes are all equally likely.) To determine the probability of obtaining a seven when a pair of dice is thrown, we determine that there are 36 equally likely outcomes and that there are six possible ways to get a sum of seven using a pair of dice: {(1,6), (2,5), (3,4), (4,3), (5,1) (6,1)}. Thus the probability of getting a seven is 6/36 or 1/6.

Tree Diagrams

A tree diagram can help determine the number of possible outcomes. To construct a tree diagram:

- Place a bold dot on the left to represent the origin or the trunk of the tree. The first set of branches represent all the possibilities for the first event to occur
- Place bold dots to mark all these possibilities
- For each dot, grow branches to represent the next set of outcomes for the second event.
- Repeat steps 2 and 3 until all events are considered.
- When you are finished, put bold dots at the end points to represent the buds of the tree.
- The buds are the number of possible outcomes. To determine the set of all possible outcomes, begin at the origin and determine the route to each bud.

Example

How many possible outcomes are there if you toss a coin and roll a die? (Have your tree diagram start with a dot, then 2 branches each ending in a dot, then 6 branches coming out of each of these dots. You should end up with a total of 12 buds.)

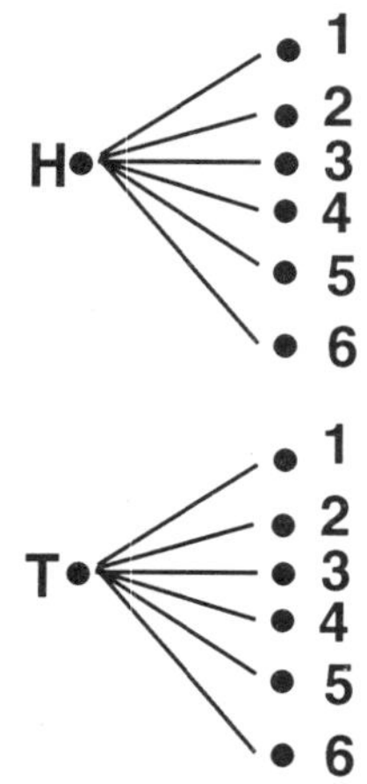

Some Basic Rules of Probability

Rules of Addition

To apply the addition rule of probability we must first determine if the events are mutually exclusive--if one event occurs none of the others can occur at the same time. For example, if you pick one card it cannot be both the queen of diamonds and the jack of hearts (unless the deck is counterfeit). If two events, A and B, are mutually exclusive, then the addition rule of probability states that: The probability of A *or* B = P(A) + P (B). (Note: we can often associate the word or with addition (+) when determining the likelihood of more than one event.)

Example

If a card is drawn from a standard deck of 52 cards, the probability of drawing a Jack or a King is: 4/52 + 4/52 = 8/52 or 2/13.

The Rule of Complement

The complement rule is used to determine the probability of an event occurring by subtracting the probability of the event not occurring from 1. If the probability of catching the flu in a certain month is .01, then the probability of not having the flu that month is 0.99 or 1 - 0.01.

Example

From a standard deck of cards, one card is drawn. The probability that this card is not a club is 1– (probability it is a club) or 1 – 1/4 = 3/4

Universal Rule of Addition

If the outcomes of an experiment are not mutually exclusive, then some events may overlap--that is, they have joint probabilities. If in determining the probability that a card drawn was either a queen or a heart, we were to simply compute the P(Q) + P(heart), we would be counting the queen of hearts twice. To correct this situation we only need subtract one of these, or the probability of drawing the queen of hearts. Thus

$$P(\text{queen or heart}) = P\,(\text{queen}) + P(\text{heart}) - P(\text{queen of hearts}).$$

In general, if events A and B are not mutually exclusive, then

$$P(A \text{ or } B) = P(A) + P(B) - P(A \text{ and } B).$$

The logic behind this formula is that when P(A) and P(B) are added, the circumstances in which A and B both occur are counted twice. To accommodate for this, P(A and B) is subtracted. Notice that if the events were mutually exclusive, then the P(A and B) would be zero. Thus, the universal rule of addition may be applied in all cases.

Example

The San Diego Tourist Commission selected a sample of 500 tourists. They found 300 went to Sea World, 100 went to the San Diego Zoo, and 50 went to both. The probability that a tourist selected at random went to either Sea World or the San Diego Zoo is: 300/500 + 100/500 – 50/500 = 450/500 = 90%.

Rules of Multiplication

To apply the rules of multiplication, it is important first to determine if the events are independent. Two events are independent if the occurrence of one does not alter the probability of the occurrence of the other. For two independent events A and B, the probability that A *and* B will occur is found by *multiplying* the two probabilities.

If a coin is tossed and a card is drawn from a standard deck of cards, then the probability of getting a head and a club is:

$$1/2 \times 1/4 = 1/8$$

In general, for two independent events, A and B, the probability of A and B happening is:

$$\mathbf{P(A \text{ and } B) = P(A) \times P(B)}$$

In other words, the probability of A and B both occurring is the product of the probability of A and the probability of B.

This rule can be applied to any number of independent events. If A, B, C, D, and E are all independent then, the probability of A, B, C, D, and E is:

$$P(A \text{ and } B \text{ and } C \text{ and } D \text{ and } E\,) = P(A) \times P(B) \times P(C) \times P(D) \times P(E)$$

Example

Drew draws a card at random out of a deck, replaces it, and then draws another card at random. What is the probability that the first card is the ace of clubs and the second card is a club (any club). Since there is only one ace of clubs in the deck, the probability of the first event is 1/52. Since there are 13 clubs in the deck and 13/52 or 1/4 of the deck is composed of clubs, the probability of the second event is 1/4. Therefore, the probability of both events occurring is 1/52 x 1/4 = 1/208 .

Conditional Probability

If two events are not independent, they are referred to, not unexpectedly, as being dependent. For example, suppose there are 100 computer chips in a box and it is known that five are defective. The probability of selecting a defective chip is 5/100 and the probability of selecting a good one is 95/100. However, if one chip is selected and then another is selected *without* returning the first one to the box, the probability that the second chip is defective depends on whether or not the first one was defective. That is, the probabilities are:

4/99 if the first chip was defective

5/99 if the first chip was not defective.

The first ratio, 4/99, is called a conditional probability because its value is conditional or dependent on whether a defective chip is chosen in the first selection from the box. A conditional probability is the probability of an event occurring given that another event has already occurred. The probability of picking two defective chips, *without replacement*, is 5/100 × 4/99, because we look at the chances of the first and second picks both being defective.

Counting

In many probability problems, the biggest obstacle is determining or counting the number of possible outcomes. There are some sophisticated and systematic ways this can be accomplished. All counting formulas are based on the fundamental theorem of counting which is very similar to the multiplication rule:

If an event A can happen in a ways and an event B can happen in b ways, then, if A and B are independent events, A *and* B can happen in a × b or ab ways.

This can be extended to any number of events. That is, if A, B, C, D, E, and F are all independent and can occur a, b, c, d, e, and f ways, then the number of ways A and B and C and D and E and F can occur is: abcdef.

Example

In a certain state a license plate consists of seven symbols composed of numbers and letters. If the first symbol is either a 1,2, or 3, and the second symbol is any letter of the alphabet except O and I, the third, fourth, fifth, and sixth can be any letter of the alphabet and the seventh is any digit except zero and one.

How many possible license plates are there?

In essence we have seven slots that need to be filled. The first slot can be filled in three ways, the second in 24 ways, the third, fourth, fifth and sixth can be filled in 26 ways, and the seventh in eight ways. Thus, using the fundamental theorem of counting, we find:
$\underline{3} \times \underline{24} \times \underline{26} \times \underline{26} \times \underline{26} \times \underline{26} \times \underline{8} = 263{,}218{,}176$

Factorial Rule (!)

The symbol ! denotes the product of whole numbers in decreasing order. For example: $6! = 6 \times 5 \times 4 \times 3 \times 2 \times 1 = 720$. If six people wish to march in a parade in which each participant lines up in back of another then there are 6! or 720 different ways to make such an arrangement. In general we have:

The Factorial Rule: n Different Elements Can be Arranged in Order n! Ways.

Example

If a candidate for president wishes to visit all 50 states, the number of different routes that he/she can take is 50!

We use the factorial rule when we wish to determine the number of ways to arrange a set of people, places or things. This rule is applied when all the members of the set will be part of the arrangement. Often we wish to arrange a subset of these members. For example, suppose 20 people show up who want to march in a demonstration, but only six can be accommodated.

We could use the fundamental theorem of counting and see that we have 6 places and can make our selection the following number of ways:

$$\underline{20} \times \underline{19} \times \underline{18} \times \underline{17} \times \underline{16} \times \underline{15}$$

Whenever we have n items to arrange and only r slots to fill, and the order in which each item is placed is significant to the problem, then such arrangements are called permutations.

Permutation Rule

The number of permutations (or arrangements in which order is important) of r items selected from n possible items is

$$_nP_r = \frac{n!}{(n-r)!}$$

Thus: $_{20}P_6 = \frac{20!}{14!}$

Example

A bank security systems device has 10 buttons and is disarmed when 6 are pushed in a proper sequence and no button can be pushed twice. To determine the number of possible codes we have:

$$_{10}P_6 = \frac{10!}{4!} = 10 \times 9 \times 8 \times 7 \times 6 \times 5 = 151{,}200$$

Another special case of the counting rule is arranging items where some are indistinguishable.

Example

The letters M I S S I S S I P P I are placed in a box and each letter is drawn at random. To determine how many different strings of letters we can draw, we look at the frequency of each letter we find:

1 **M**, 4 **I**'s, 4 **S**'s, 2 **P**'s

We have 11 slots to fill. If each letter WAS distinguishable we would have 11! possibilities. However, since each of the repeating letters is not distinguishable, we need to divide out all the replications. Thus,

$$\frac{11!}{4! \times 4! \times 2!} = 34{,}650$$

Combinations

Many situations involve the selection of items where order does not count. For example, suppose that we have six people: Ali, Barbara, Carlos, Dave, Edna, and Frank, who wish to be on a committee with only two available places. It does not matter in what order these people are selected; that is, if we select Ali first then Barbara or Barbara first and then Ali, it wouldn't matter since they both would be on the committee. If we list all possible ways to fill these positions we get: {A,B}, {A,C}, {A,D}, {A,E} {A,F}, {B,C}, {B, D}, {B,E}, {B,F}, {C,D},{C,E}, {C,F}, {D,E}, {D,F}, {E,F} . When different orderings of the same items are not counted, we have a combination situation and can apply the following rule:

Combination Rule

The number of combinations of r items selected from n available items is:

$$_nC_r = \frac{n!}{(n-r)!r!}$$

$$_6C_2 = \frac{6!}{(4!)(2!)} = 15$$

Notice how the sum of the factorials in the denominator equals the factorial in the numerator.

Example

Juanita claims she has ESP, so she doesn't have to study for tests. Lu decides to do an experiment and constructs a 12-question multiple choice test on sewing, which Juanita knows nothing about. Each question has three choices and one correct answer. Juanita gets nine out of 12 correct. She claims that her ESP isn't perfect but is better than guessing. How likely is she to get nine out of 12 by just guessing?
Let C = correct answer and I = incorrect answer.
First we can determine the number of ways to arrange 9 C's and 3 I's:

$$\frac{12!}{9!3!} = 220$$

One of these ways is: CCCCCIICICCC . The probability of this particular arrangement is $1/3 \times 1/3 \times 1/3 \times 1/3 \times 1/3 \times 2/3 \text{ x } 2/3 \times 1/3 \times 2/3 \times 1/3 \times 1/3 \times 1/3 = 8/312$
Thus, the probability of just guessing and getting exactly nine correct is:
$220\ (8/3^{12})$, which is less than 1%. However, this does not include getting more than nine correct. To determine the latter we would need to calculate the probability of getting 10, 11, or 12 correct. There are much easier ways to do this to be found in the study of what is known as *binomial distributions*.

IMPORTANT FORMULAS	
$0 \le P(A) \le 1$	The probability of any event
P(A or B) = P(A) + P(B)	If A and B are mutually exclusive
P(A or B) = P(A) + P(B) - P (A and B)	If A and B are not mutually exclusive
P(A') = 1 - P(A)	Complement of event A
P(A) + P(A') = 1	An event and its complement covers it all
P(A and B) = P(A) × P (B)	If A and B are independent
P (A and B) = P (A) × P (B| A)	If A and B are dependent
$m \times n$	The number of ways two independent events can occur
$n!$	The number of ways that n different items can be arranged
${}_nP_r = \frac{n!}{(n-r)!}$	The number of permutations when n items are taken r at a time
${}_nC_r = \frac{n!}{(n-r)!r!}$	The number of combinations when n items are taken r at a time

WHAT'S STAT YOU SAY?

Despite what you might have been told, or have experienced, statistics is not SADISTICS. This misconception probably comes from some politicians who use statistics the way that a drunk uses a lamppost--more for support than illumination

To help you understand what statistics IS and what it can and cannot do I offer you the following 8 statements:

1. The purpose of statistics is to collect, organize, and analyze data, and then interpret the results and make predictions. You can perhaps remember the Acronym COAIP.
2. One relies on statistics to determine "how close" to what one anticipates would happen actually did happen.
3. When one uses statistics they are comparing a numerical result to a number that is reflective of a chance happening and determining how significant the differences between these numbers are. [This is why statistics relies heavily on probability].
4. Statistics can be used to predict, but that these predictions are NOT certainties.
5. Hypotheses are educated guesses that are derived by logical analysis using induction or deduction from ones knowledge of the problem and from the purpose for conducting a study. They can range from very general statements to highly specific ones. Most research studies focus on the proving or the disproving of hypotheses.
6. In hypothesis testing, descriptive numbers are standardized so that they can be compared to fixed values (found in tables or in computer programs) that indicate how "unusual" it is to obtain the data collected.
7. Certain conditions are necessary prior to initiating a statistical test.
8. Once data are standardized and the significance level determined, a statistical test can be performed to analyze the data and possibly make inferences about an entire population (universe).

The fact that conclusions may be incorrect separates statistics from most other branches of mathematics. If a FAIR coin is tossed ten times and ten heads appear, the statistician would incorrectly report that the coin is biased. This conclusion, however, is not certain. It is only a "likely conclusion," reflecting the very low probability of getting ten heads in ten tosses.

Statistics are ubiquitous, they are prevalent in nearly every aspect of our lives. Manufacturers use techniques to control processes and ensure quality for their customers. Marketers use demographics to target segments likely to have interest in their products. The Government tracks key indicators to predict and to effect future economic trends. Statistics are used to decide what television shows, medicines, and educational programs will be made available to us. An understanding of the concepts utilized by these agencies is imperative if we wish to exercise influence in these areas.

An **average** is a measure of central tendency. There are many types of averages (mean, median, mode, midrange) but the *arithmetic mean average*, or simply the *mean* is often the most reliable measure of central tendency since it is the one that is the most "sensitive" to all the data.

Although the mean does consider the relative "weight" of all data, the mean does not, by itself, give us a picture of how the data is *dispersed.* To accomplish this, the *standard deviation* or the *variance* (which is the standard deviation squared) is often reported.

If the standard deviation is relatively large compared to the mean, a wide dispersion of numbers is indicated. Conversely, a small standard deviation indicates that the numbers cluster around the mean. Thus, it is generally expected that researchers report the mean and standard deviation when presenting their findings.

There is an empirical rule that applies to data having a histogram that is approximately bell shaped.

- About 68% of all scores fall within one standard deviation of the mean.
- About 95% of all scores fall within two standard deviations of the mean.
- About 99% of all scores fall within three standard deviations of the mean. That is, the mean I.Q. on the Stanford-Binet Test is 100 and the standard deviation is 15. Thus, 68% of those who have taken the test have I.Q.'s between 85 and 115. This is often referred to as "normal." Those with I.Q.'s between 115 and 130 are considered "above average," those with I.Q.'s between 130 and 145 are considered "genius," above 145 is "brilliant."

A *standard score*, or *z score*, is the number of standard deviations that a given value is above or below the mean. Z values usually range from –3 to 3. In general, if two z values differ by *less than 1.65* from each other, that difference is *not* considered statistically significant.

A *standard normal probability distribution is a Bell-Shaped* curve (also called a *Gaussian curve* in honor of its discoverer Karl Gauss) with the area under the curve equal to 1, the mean, or middle value 0 and the standard deviation, the place where it starts to bend, equal to 1 on the right and –1 on the left. It serves as a graphical representation of z values and what percentage of the population would be assigned that value. By computing a z value you are actually converting your mean to a mean of 0 and your standard deviation to a standard deviation of 1.

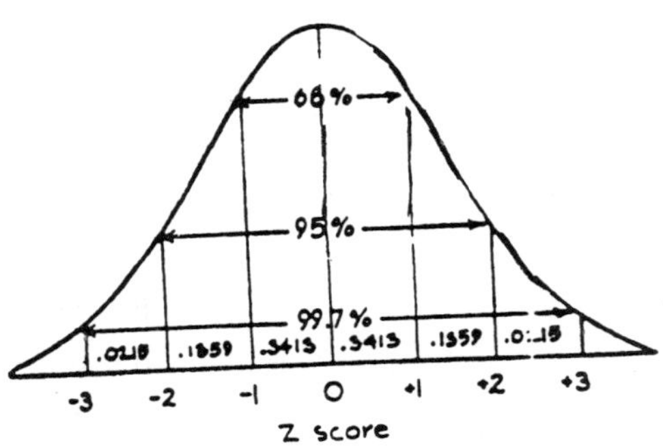

CALCULATING CALCULUS

The Calculus

Göttfied Wilhelm Leibniz (1646-1716) is best known as a mathematician, but he was a master of many trades: law, religion, politics, history, philosophy, and science. By age 20 he had earned his doctoral degree. His greatest interest was to develop mathematics that would be interdisciplinary. He invented binary numbers and related them to his religious beliefs. He is believed by many (in particular German mathematicians) to have invented the calculus . . . or did he?

Sir Isaac Newton (1642-1727) is considered the greatest English mathematician and scientist who ever lived. He discovered mathematical and physical laws that govern the universe. At 21 he is also believed (in particular by English mathematicians) to have discovered the calculus, which he called "flusions." Some have suggested that both Leibniz and Newton had studied with the same professor--William Barrow--so they might both have borrowed from Barrow!

What separates calculus from other branches of mathematics is the idea of limits. When we study the limit at a certain point we can determine what happens as we get "very close" to that point without actually getting there. In the infinite sequence below where each term is half the previous term,

8, 4, 2, 1, 1/2, 1/4, 1/8 . . .

the limit of this sequence is 0, yet we can never quite get there. However, we can get as "close as we want to" by going far enough out in the sequence.

There are two main branches of calculus: *differential* and *integral*. In differential calculus we are concerned mostly with rates that are changing. In algebra we study "average rates of change," but in calculus we study "instantaneous rates of change," or the rate of change at a precise moment in time. Such a rate is known as a *derivative*. The geometric interpretation of a derivative is the slope of a line tangent to a curve at a point on the curve. To find the slope of a line in algebra we use two points. In calculus we can find the slope of the line at one point with the aid of the limit concept. In fact, calculus can be considered elementary math (algebra, geometry, and trigonometry) enhanced by the limit process.

The slope of line AB is the derivative of the curve at B.

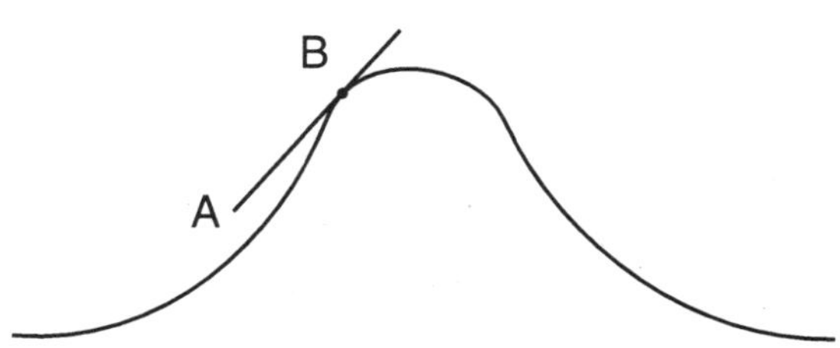

The derivative may also be thought of as the speed of a moving particle or the rate of change of one variable with respect to another.

For the function $y = f(x) = x^2$, the derivative function, y' or $f'(x)$ or $dy/dx = 2x$. This tells us that the slope of the line tangent to the curve at the point where $x = 1$, or $f'(1) = 2$, whereas the slope of the line tangent to the curve at $x = -1$ or $f'(-1) = -2$, and the slope of the line tangent to the curve at $x = 0$ is $f'(0) = 0$. There are general and special formulas to find the derivative of functions.

In *integral calculus* we are concerned with the geometric problem of finding the area under a curve. Although this might seem pretty trivial, just think about how we find the area inside of a circle. This exercise involves dealing with the infinite irrational number, π! If the derivative of a function at each point is known, then through integral calculus we can obtain the function itself. The relationship between differential and integral calculus is similar to the relationship between multiplication and division.

The calculus is merely rules associated with these two new operations--differentiating and integrating, which are needed when longer-established mathematical methods fail to accomplish the task at hand. Broadly speaking, all problems which involve continuous changes, be they changes of positions, time, size, shape, motion, money, pitch, volume, and so forth, call for the use of this calculating technique we call *the calculus*. Just as arithmetic is well suited for numerical calculations, so the calculus is the mathematical technique well fitted for calculations involving quantities that are varying continuously in magnitude such as the speed of a train, the metabolism of a drug, the growth (or shrinkage) of a tumor, the profit from an investment, or the loudness of a singer's voice. Once the concepts underlying the twin operations of differentiation and integration are grasped, the simplicity and power of this method of calculation can be understood and mastered.

CHAOS THEORY AND FRACTAL GEOMETRY–BOUNDED MADNESS

In Greek mythology, chaos was the unorganized state or void from which all things arose. Proceeding from time, Chaos eventually formed a huge egg from which there issued Heaven, Earth, and Eros (love). According to Hesiod's "Theogony," *chaos preceded the origin not only of the world, but also of the gods.*

Introduction

Chaos theory and fractal geometry radically question conventional comprehension of equilibrium, harmony, and order, in nature as well as in other contexts, and offers a holistic and integral model which can encompass the true complexity of nature for the first time. They offer us a much more adequate understanding of ecology and climatic developments, and thus the hope of being able to more effectively address our seemingly overwhelming global challenges.

The Basics of Chaos Theory

A New Lens to View the World

Never before have mathematical insights, usually regarded with confusion, found such rapid acceptance or generated so much excitement in the public mind. In the famed movie and book *Jurassic Park*, followers of chaos theory are described as the new generation of mathematicians who are openly interested in how the real world works (Crichton, 1992). In the popular mind, the study of chaos and fractal geometry has brought mathematics into the twenty-first century by offering a unique vision of the dynamics of mathematics, its many interconnecting links, and its ability to correct outmoded conceptions of the world. Prior to the entrance of this new vision, the illusion persisted that the world on the whole functioned like a huge clockwork mechanism whose laws were only waiting to be deciphered step by step. What is now accepted is that strict determinism and apparently accidental development are not mutually exclusive, but rather coexist in nature.

According to Gleick, the basic idea of Western science is that you don't have to take into account the falling of a leaf on some planet in another galaxy when you are trying to account for the motion of a billiard ball on a pool table on earth. Very small influences are neglected. There's a convergence in the way things work, and arbitrarily small influences don't blow up to have arbitrarily large effects (Gleick, 1987, p. 15). Chaos theory is proving Western science wrong. There are many components, implications, and conjectures inherent in chaos theory. Although an exact definition of chaos theory is ephemeral (on par with an *exact* definition of happiness) Figure 1 is an attempt to give the reader an overview of the spirit and focus of this burgeoning discipline.

CHAOS THEORY & FRACTAL GEOMETRY

WHAT IS CHAOS?

The Study of Nonlinear Interdependent Systems that Exhibit Sensitive Dependence on Initial Conditions.
A Far From Equilibrium System Which Can Undergo Self-Organization.
A Random Vacillation Within a Certain Range.
A Constrained Type of Randomness.
A Dynamical System Which:
Operates under a wide variety of conditions,
is based on feedback or iterations, and
makes exact precision near impossible.
A Process Which can Be Explained by Nonlinear, Reiterating Equations.
The Morphology of Amorphous.
The term Chaos was coined by T.Y. Li and J.A. Yorke in 1975.

WHAT IS A FRACTAL?

A Geometric Pattern Left Behind By Chaotic Actions.
A Shape of Varying Size with Similar Shapes at Finer Detail.
A Form in Which Each Section can Give Rise to a Similar Pattern which May be Slightly Different Than the Original.
The Dimension of a Figure which is between 1 and 2, 2 and 3, 3 and 4, etc.; thus it is not integral. A Fractal Dimension is defined by many algorithms e.g. the Correlation Dimension, The Information Dimension, and the Capacity or Hausdorf Dimension.
The Fractal Dimension of a set is a Number which tells how densely the set occupies The Metric Space in which it lies.
A Measure That Can Describe The Complexity of Sensory Experiences.
The Geometric Shape of a Strange-Attractor.
The term Fractal was coined by Benoit Mandelbrot in 1975.

SOME EXAMPLES of the Relationship BETWEEN CHAOS & FRACTALS

CHAOTIC PROCESS	FRACTAL
Seashore	Coastline
Atmospheric Pressure	Clouds
Earthquakes	Rock Formation
Embryonic Development	Human Body

WHAT CAN CHAOS THEORY EXPLAIN?

Weather, population, stock market, heart attacks, earthquakes, how the brain works, Dow-Jones averages, chemical reactions, investments, ecology, metabolism...

Figure 1

In classical science, there is an ideal goal of total physical predictability, according to Nobel Prize-winning chemist Ilya Prigogine (1980). However, "Total predictability was never more than a theoretical possibility. Yet in some sense this unlimited predictability was an essential element of the scientific picture of the physical world" (p. 214).

Since total predictability requires that a system not manifest sensitive dependence on initial conditions, a system characterized by point values that evolve along a unique trajectory will be unpredictable if small differences in initial conditions lead to widely separated trajectories later on. Since no measurement can be made with unlimited accuracy, a chaotic system will be unpredictable even though it may in other ways be deterministic. Thus, total physical predictability appears to be an illegitimate ideal exposed by chaos theory (Kellert, 1993).

Why the Interest in Chaos Theory?

Certain patterns of structure recur again and again. Fractal principles have given us valuable insight into how nature creates the shapes we observe. Mountains, rivers, coastlines, vegetables, lungs, circulatory systems--all are fractal, replicating a dominant pattern at several smaller and larger levels of scale.

Left to develop and change randomly with only a few simple guidelines, nature creates the complexity and harmony of form we see everywhere. The theories of chaos can be viewed as the life source of our creative power and a new tool in understanding and explaining things we could not explain and understand before. These concepts have contributed to remarkable progress in the deeper understanding of phenomena such as turbulence, cloud formation, fibrillation's of the heart, laser instabilities, population dynamics, climate irregularities, brain function anomalies, the stock market, etc. Those who have come to see mathematics as a dead subject in which everything has already been discovered will gain a very different view of mathematics when they realize the fact that chaos theory, dynamical systems and fractals are not completely understood.

Chaos theory is an example of a complex dynamical system. According to Waldrop (1992), in the natural world such systems include brains, immune systems, ecologies, cells, developing embryos, and ant colonies. In the human world such systems include cultural and social systems such as political parties or scientific communities. Each of these systems is a network of many "agents" acting at the same time. Each agent finds itself in an environment produced by its interactions with the other agents in the system. Each is constantly acting and reacting to what the other agents are doing. If there is to be any coherent behavior in the system it has to arise from competition and cooperation among the agents themselves. To be adaptive, a complex system must constantly make predictions based on its various internal models of the worlds--its implicit or explicit assumptions about the way things are. Typically, each system has many niches, and is always in transition.

Some Main Ideas and Mathematics Behind Chaos Theory

The Butterfly Effect. A butterfly flaps its wings over St. Paul and sets in motion a storm in Anchorage. With the next flap, nothing of meteorological significance happens. We see how a small input can have large effect in the adage, "The straw that broke the camel's back." The slightest variation can sometimes have an explosive result. Even the smallest vagueness in conversation can eventually blossom into utterly open ambiguity.

Sensitive Dependence on Initial Conditions. Sometimes an extremely small input can have an extremely large result (see the Butterfly Effect). A dynamical system which exhibits this property will produce markedly different solutions for two specified initial states that are initially very close together. Given any specified initial condition, there is another set of initial conditions infinitesimally close to it that will diverge from it by some required distance, given enough time, thanks to the process of feedback.

Feedback. Principles fundamental to all sciences. Feedback occurs when the same operation is carried out repeatedly; the output of one iteration becomes the input for the next one. In home furnaces, when a room cools down below a temperature set on a thermostat, the thermostat responds by switching on the furnace, and the activity of the furnace affects the thermostat. The furnace and the thermostat are bound in what is technically called a negative feedback loop (which regulates). This is also seen in the relationships between predator and prey or in the checks and balances of the U.S. Constitution. Examples of positive feedback (which amplifies) include ear-splitting screeches produced in a public address system when the microphone is placed too close to the loudspeaker. The screeching sound is the result of an amplifying process in which the output of one stage becomes the input of another. In algebra and geometry, feedback is called iteration. Consider $f(x) = 2x + 1$, and let $x = 1$, $x = 2$. If we find f(1) we would substitute 1 for x and get 3. If we find f(f(1) this would give us f(3), which is seven. If we keep going, f(f(f(1))), we get larger and larger values. However, when we iterate the function $f(x) = 4x(1 - x)$, and start with $x = 0.8$ and find the tenth value through iteration, this gives us something totally different than if we started with $x = 0.79$ or 0.81 and did the same number of iterations.

Attractors. States which most dynamical complex systems eventually settle into until some major perturbation or change occurs. No matter which initial input we choose in a system, we obtain a sequence of images that always tends toward one and the same final image. An attractor can be thought of as a region which exerts a "magnetic" appeal for a system, seemingly pulling the system towards it. In cultural evolution, attractors have been bands, tribes, chiefdoms, and nation-states.

Just for fun, iterate the function $f(x) = x^2$ for the following values: 1, –1, 0.5, and 2. What did you find?

DELIGHTFULLY DISCRETE MATHEMATICS

Discrete mathematics is a rapidly growing and increasingly used area of mathematics with many practical and relevant applications. Because it is grounded in real world problems, discrete math lends itself easily to implementing the recommendations from *The National Council of Teachers of Mathematics - Curriculum and Evaluation Standards* (NCTM Standards, 1989). Discrete mathematics is an exciting new mathematics course that has been widely recommended for inclusion in K-12 and undergraduate education. You can use discrete mathematics to create efficient computer networks, optimally assign frequencies to cellular phones, track pollution in an ecosystem, fairly rank tournaments, choose the best candidate in presidential primaries, efficiently schedule large projects, plan optimal snow-plowing routes, solve puzzles, color maps, and have fun with mathematics!

Discrete mathematics is not new. Many of its techniques emerged in the eighteenth century. However, it is only since the middle of the nineteenth century that it has been recognized as a separate branch of mathematics. Discrete math helps make mathematical concepts come alive. It's an excellent tool for improving reasoning and problem-solving skills, and is appropriate for students at all levels. Problems in discrete mathematics are often simply stated and accessible to students with little mathematical background, including those who have not mastered algebra. However, many problems are complex, leading students to new insights and knowledge. Questions that seem simple often give rise to related questions that are quite challenging.

Because discrete mathematics deals with complex problem solving, it has widespread uses in a variety of fields. Many current applications are in areas such as networking, telecommunications, computer design, robotics, social choice theory, and operations research involve discrete mathematics. It utilizes concepts found in the study of matrices, set theory, linear programming, chaos theory, graph theory, probability and statistics, modular arithmetic, bases, number theory, Pascal's Triangle and more.

There are three main categories that discrete math problems fall into.

1. *Existence problems* that deal with problems that may or may not have a solution. For example, if you are trying to arrange a tournament, is it possible for each team to play every other team and have no team play three games in a day?
2. *Counting problems* that investigate how many solutions are possible. For example, as more and more people use cellular phones, faxes, and e-mail, there is a need for more phone numbers and area codes. Discrete math helps determine the number of phone numbers needed for each geographical region.
3. *Optimization problems* that focus on finding the best solution to a problem when there are several possibilities. NASA deals with these types of problems in practically all aspects of launching a space shuttle. Some other types of problems in discrete mathematics include:

- What is the quickest way to sort a list of names alphabetically?
- What way of connecting a number of sites into a telephone network requires the least amount of cable?
- What version of a lottery gives the best odds?

- How can representatives be apportioned fairly among the states using current census information?

Set Theory

A set is simply a group or collection of things. Objects are usually grouped because they have some common feature or characteristic. A set of books may all be encyclopedias; a set of fast food places may all be called Burger World. Items of a set are called members or elements.

The science of sets tells us how to combine sets and how to compare sets to determine relationships. Studying discrete math, as well as most advanced mathematics, becomes simpler by using the language of sets.

Often, pictures (Venn diagrams) are used to represent sets. Traditionally, the universal set (a large set which includes all items under discussion and sometimes even more than that), is represented by a rectangular box. A subset of the universal set is usually represented by circles. Intersecting circles means there are members in common.

Example

A survey of 200 freshman revealed that
80 students were taking History
89 students were taking Physics
64 students were taking Math
45 were taking both History and Physics
23 were taking both History and Math
35 were taking both Math and Physics
3 were taking History, Physics, and Math

1. How many students were taking History but not the Physics nor Math?
2. How many students were taking History and Math but not Physics?
3. How many students surveyed were not taking any of these courses?

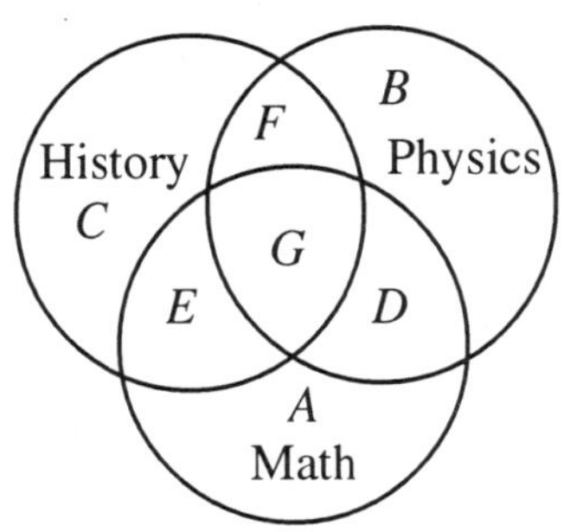

Answer Use the three intersecting circles above. Identify the circles as: History, Physics, and Math. Start with the region that is the intersection of all three circles, G. Put "3" inside of this region. Next, go to the intersection of History and Physics. This will include the "3" and must total 45; thus, the other region, F, has 42 members. Place 42 in this region. The answers are: 1.15, 2. 20, and 3. 67

Number Patterns

The calendar system is a modular seven system. This means that every seven days we are back to where we started from. It is easy to determine what day of the week it will be from a given day by simply dividing the number of days by 7, and looking at the remainder.

Example If today is Saturday, what day of the week will it be in nine days from now? 200 days from now? In 1,000 days from now?

Answer Let us use the following coding system:

0 = Saturday, 1= Sunday, 2 = Monday, 3 = Tuesday, 4 = Wednesday,
5 = Thursday, 6 = Friday.

If today is Saturday, then in nine days it will be Monday since 9/7 leaves us with a remainder of 2 and 2 = Monday in our coding system. Thus, 200/7 gives us a quotient of 7 and a remainder of 4. That means that the day will be Wednesday and it will come seven weeks from now. In 1000 days it will be Friday since we have a remainder of 6 and a quotient of 142.

Pascal's Triangle

1
1 1
1 2 1
1 3 3 1
1 4 6 4 1
1 5 10 10 5 1
1 6 15 20 15 6 1
1 7 21 35 35 21 7 1

Numerical patterns using iteration are an important component of mathematical thinking and discrete mathematics. The building of *Pascal's Triangle* has many applications in the real world and offers an excellent means of discovering mathematical relations. The first eight rows of Pascal's Triangle are constructed above.

Blaisé Pascal (1623-1662) was a great French mathematician and scientist. By the age of 20 he had already built 10 machines that could add integers--precursors of modern computers. What is known as the arithmetic triangle, or Pascal's Triangle (see above) was not, however, Pascal's discovery. A Chinese version of Pascal's Triangle had already been published in 1303. The first and last number of each row is 1. To determine the occupant of each cell between the first and last, simply add the two numbers (x and y) in the row above and place the sum (x + y) in the next row in the middle of these two numbers. Thus, the 3 in row 4 is the sum of 2 + 1. There are many hidden patterns in Pascal's Triangle. To name a few, the sum of each row is always a binary number (a multiple of 2), the second diagonal is the counting numbers, and the third diagonal is the set of triangular numbers.

What Pascal did was to use the triangle to solve many problems in algebra and probability theory. Pascal's Triangle can be used to easily find the solution to many problems.

Example Expand (x + y) 5.

Answer The row members make up the coefficients of the binomial expansion]. The sum of the exponents will equal 5 for each term, thus:

$1x^5y^0 + + 5x^4y^1 + 10x^3y^2 + 10x^2y^2 + 5x^1y^4 + 1x^0y^5$ or

$x^5 + + 5x^4y + 10x^3y^2 + 10x^2y^2 + 5xy^4 + y^5$

Example How many committees of four can be formed from a group of seven people?

Answer Go to the row beginning 1, 7, 21 . . . The first term of the row is the number of ways to form a committee of zero members from 7; (1), the next term, is the number of ways to choose one member from 7; (7), the next, is the number of ways to choose 2 people from 7; (21), the next term, is the number of ways to choose 3 people from 7; (35), the next term, is the number of ways to choose four people from a group of 7; (35). In general, to choose r people from n, go to the 1,n, row, then count from 0 to r.

Fibonacci Series

Fibonacci started with 1, and then repeated it--1, 1. Then he added the two: 1, 1, 2. And he kept on adding, always the concluding pair: 1, 1, 2, 3, 5, 8, 13, 21, and so on. He came up with the genealogy of the male bee--which has only one parent, the queen (who herself has two parents), and the answer to the question, "How many pairs of rabbits can be produced from a single pair of rabbits, if every month every pair begins a new pair, which from the second month on is itself productive (assuming no rabbits die) and each consists of a male and female?

In general, a Fibonacci series is a succession of numbers in which each term is the sum of the two terms immediately preceding it. It was named for its discoverer, Leonardo Fibonacci (or Leonardo of Pisa) in 1202. The numbers in the sequence above, 2, 3, 5, 8, 13, 21, 34, 55 may also be referred to as Fibonacci numbers. Fibonacci series have proved useful in number theory, geometry, the theory of continued fractions, and genetics.

The sequence of numbers generated can be found in many natural forms. The ratio of scales distributed in opposing spirals around a pine cone is 5:8, and bumps around a pineapple, 8:13; these are adjacent Fibonacci pairs. It has been known for several hundred years that most flowers have a Fibonacci number of petals (e.g., there are 55 in many daisies).

The ratio of two adjacent Fibonacci numbers, after 3, is roughly 1: 1.618, which is the ratio behind the celebrated Golden Section--in mathematics the division of a line segment into two pairs such that the whole segment is to the larger part as the larger part is to the smaller. The Golden Rectangle, whose length and width are the two parts of a line segment divided Golden Section-style, is so pleasing to the eye that artists claim for it to possess both static unity and dynamic variety. One classic example can be seen on the facade of the Parthenon. In addition, many aesthetically pleasing paintings are examples of the Golden Rectangle.

MATH POWERS

The following problems are designed to further increase your mastery and enjoyment of mathematics and help prepare you for future mathematical endeavors. These problems are considered "open-ended" and offer you an opportunity to pose your own questions, make your own mathematical connections, and explore mathematics on your own or with your friends, colleagues or family members. Since there are many possible answers, Math Advancement will not provide you with any solutions, but will offer you concepts that might help you get started on finding solutions. It is the hope of the author that you will see the relevance of the mathematics that you are learning in Math Advancement to these thought provoking questions. Once you make that connection, you will obtain Math Power which can ignite future success in mathematics.

1. Use the order of operations and any mathematical operations that you are familiar with to arrange the set of numbers below to obtain a value of 21. See how many different ways you can accomplish this. [concepts 1 and 2]

   a) 8, 7, 6, 5 b) 4, 5, 6, 7 c) –2, 5, –5, –2

2. Circle O is inscribed in square ABCD. If you know the diameter of the circle is 10, what else do you know? [concepts 11 and 12]

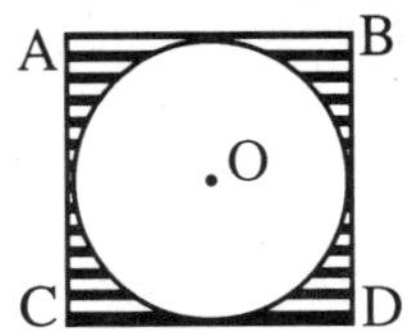

3. If line L is parallel to line M and each intersect line N as shown and the measure of angle x is 40 degrees, what else do you know? [concept 10]

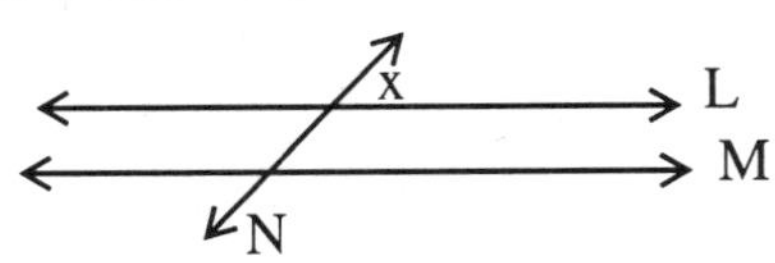

4. If there is a 20% off sale at Simon Mathtique, and you paid \$16 for a calculator (before tax), what else do you know? [concept 5]

5. Use the distribute property to find several ways to multiply 9×99. [concept 8]

6. If 2.54 cm equals 1 inch, what else do you know? [concept 3]

7. If the average of 5 numbers is 90, what else do you know? [concept 9]

8. Triangle ABC is equilateral and the length of AB is 16 cm. By connecting the midpoints of each side of triangle ABC, 4 congruent triangles were formed and the middle one was shaded in. Repeat this process by connecting the midpoints of each of the non-shaded triangles and shading in the middle triangle for each set. How many congruent triangles are now shaded? Unshaded? What is the total area and perimeter of the shaded regions? non shaded regions?

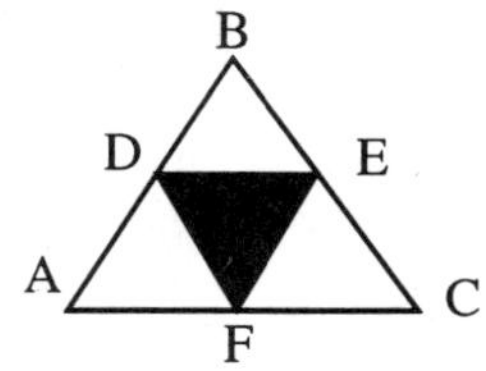

9. Suppose that the average (arithmetic mean) cost of college today is $12,000 per year with an inflation rate of 8%. The "rule of 72" says that at this rate, the cost will double every 9 years. [The rule of 72 says the principal will double when the per cent $\times$ years is about 72]. If the inflation rate stays at 8%, how much would it cost to put your children or grandchildren through college in four years?

WHAT'S YOUR PREFERENCE?

One of the wonderful things about mathematics is that there are so many ways to go from what you know (given) to where you want to go (prove). Let's take a look at four different ways to multiply the binomials: $(X + 4)(X + 5)$

Method One

Using the distributive property, let's distribute $(X+ 4)$ over each term in $(X+ 5)$
This gives us:

$$(X+4)X + (X + 4)5.$$

Now let's distribute both the X and the 5 over $(X +4)$. This gives us:

$$X^2 + 4X + 5X+ 20 \text{ or } X^2 + 9X + 20$$

Method Two

Since 14 is 10 + 4 and 15 = 10 + 5, we can look at the way we multiply $(X + 4)(X + 5)$ by comparing it to the way we multiply (14) (15). This gives us:

$(X + 4)$		14	$(10 + 4)$
$(X + 5)$		$\times$ 15	$(10 + 5)$
$5X + 20$		70	(or $50 + 20$)
$X^2 + 4X$		1 4	(or $100 + 40$)
$X^2 + 9X + 20$		2 10	(or $100 + 50 + 40 + 20$)

Method Three

Notice that each term of $X + 4$ gets to multiply each term of $X+ 5$ once and only once. The acronym FOIL can be used to keep track of all this multiplying. **F**: First terms; **O**: Outside terms; **I**: Inside terms; **L**: Last terms:

$$(X + 4)(X + 5) = X^2 + 5X + 4X + 20 = X^2 + 9X + 20$$

F O I L

Method Four

To find the area of a rectangle we multiply: (length) X (width). If we have a square (which is a special type of rectangle) that measures X on each side, then the area of that square is X^2. If we had a square that measured 1 on each side, then the are of that square would be 1 x 1 or 1. If we had a rectangle that measured X on one side and 1 on the other, then it would have an area of 1(X) or X.

In the figure on the right, we have a rectangle that is $(X + 4)$ by $(X + 5)$. To find the area we can take an inventory of the area of the figures inside. Notice that we have one square with area X^2, 9 rectangles with area X and 20 squares with area 1, or $= X^2 + 9X + 20$

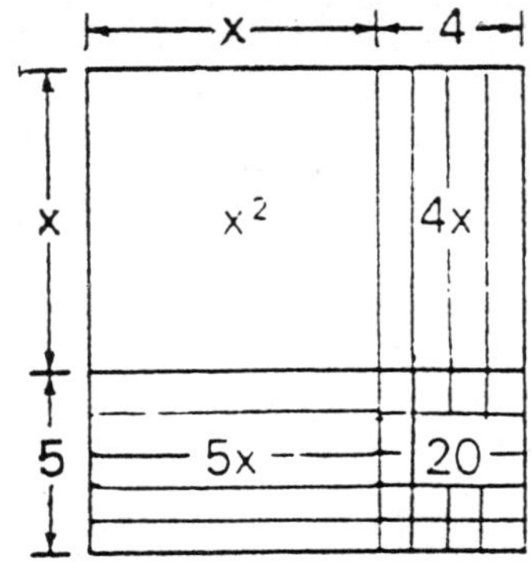

WORDS AND IDEAS FOR MATH STARS

Directions: Match the numbers in COLUMN II that are described in COLUMN I (Answers on the bottom)

	COLUMN I		COLUMN II
1.	Prime Numbers	A.	60
2.	The set of Integers	B.	{. . . –3, –2, –1, 0, 1, 2, . . .}
3.	The greatest common factor of 8, 12, and 20	C.	{2, 3, 5, 7, 11, 13}
4.	The lowest common multiple of 15 and 20	D.	1 : 4
5.	The probability of picking a spade from a standard deck of cards	E.	1,000,000
6.	The odds of picking a spade from a standard deck of cards	F.	1,000,000,000
7.	The perimeter of a square 5 cm by 5 cm (in cm)	G.	{2, 4, 8, 16, 32, 64, . . .}
8.	The area of a square 5 cm by 5 cm (in square cm)	H.	25
9.	A million	I.	125
10.	A billion	J.	20
11.	Powers of 2 (binary numbers)	K.	4
12.	Factors of 12	L.	20.0
13.	19.998 rounded to the nearest tenth	M.	20.00
14.	19.9997 rounded to the nearest hundredth	N.	{1, 2, 3, 4, 6, 12}
15.	The volume of a cube 5 cm by 5 cm by 5 cm (in cubic cm)	O.	1 : 3

1. C 2. B 3. K 4. A 5. D 6. O 7. J 8. H 9. E 10. F 11. G 12. N 13. L 14. M 15. I

EXERCISES FOR TEST TAKERS

Some simple exercises, all done in your seat, can make a long test more tolerable by stimulating circulation and relaxing muscles. The exercises can be done so discreetly that your fellow test takers may not realize what you are doing.

As you do each exercise, remember to sit up straight and breathe evenly and deeply. Then do as "Simon Says:"

1. Lean head forward as far as possible. Feel a stretch down the back of your neck. Lean head to the side (keep face front) and feel the stretch down the side of the neck. Lean head to the other side.

2. Let head drop back (jaw relaxed) and look up at the ceiling. Arch upper back as you look up. Feel stretch under your chin.,

3. Hug yourself (right hand on the left shoulder, left hand on the right shoulder). Lean forward, cave in upper chest and pull your shoulders forward. Feel stretch down the back of the neck and between the shoulder blades.

4. Relax your shoulders and arms. Make a circle with one shoulder at a time by moving it forward, raising it up, pushing it back and down. Then move it back reversing the exercises.

5. Press elbows down onto the desk and press shoulders down as hard as you can. Feel stretch across top of shoulders.

6. Sit up straight and pick one foot up off the floor by raising your whole leg about an inch off your chair. While maintaining this position, make circles in the air with the raised foot by rotating your ankle. Circle your foot to the right eight times and to the left eight times. Place the foot back on the floor and repeat with your other foot.

Do these exercises intermittently during the test. After answering five questions you should be ready for a little relaxation. Try these exercises at home while doing your homework before you take a major test. Find out which ones work the best for YOU!! Remember to breathe evenly during all the exercises.

HELPFUL HINTS

✓ **The Night/Day Before the Test.** Try to leave 24 hours between studying for a big exam and taking that exam. A pasta dinner might help you sleep better and help you wake up feeling refreshed. Strive to get your optimal amount of sleep: 7-10 hours. Try watching a comedy or reading an inspirational book the night before the test.

✓ **The Morning of the Test.** A breakfast for test Champions consists of: protein, carbohydrates and potassium. If you have not had breakfast in years, you should start experimenting with different types of breakfasts before the morning of your test. Plan to get to the test early so that you can be comfortable with the surroundings.

✓ **What to Wear to the Test.** Dress for success. Wear something that is comfortable and flattering. If nothing suits that bill then this is an excellent opportunity to go shopping for a test outfit or to exchange an outfit with a friend. Cotton and layered clothing might help you feel more relaxed.

✓ **What to Bring to the Test.** If you will be traveling to the test by car, make a mental note of the materials necessary for the test (pencils, watches, admission card, ID's, water, etc.). You can then visualize placing these objects on your cars dashboard, or on your shoe if you will not be driving your car to the test. You might wish to bring a bottle of water and/or a box of raisins to the test to provide you with some extra fuel and lubrication during the test. Research suggests that listening to Mozart or instrumental complex music right before your exam might enhance your memory.

✓ **How to Relax During the Test.** Negative images and messages could pop into your head during a test. Tests sometimes cultivate the remembrance of words and actions that diminish our sense of worth. Anytime you catch one of those negative thoughts during a test, it is important to **stop**, put down your pencil, take a deep breath, think of something positive, do a stretching exercise and then return to the test. Asking yourself about the question at hand, or going on to the next question can also help eliminate the negative message. There is some indication that the scent of vanilla is very calming.

✓ **How to Get the Right Answer onto the Answer Sheet.** Most BIG tests do not allow scrap paper - you are told, instead, to write on the test. Write all your answers in the test booklet and then when you are ready to turn to another page, that would be an excellent time to transfer all the answers on that page onto the answer sheet. Once you have accomplished that, do a quick stretching exercise.

✓ **How to Pace Yourself.** When you read a question for the first time, use your intuition to answer that question. Don't look to justify your answer at this point. IF you just "know" the answer, that is fantastic! The next time you see a question, look for a trigger word to jolt your memory bank or try relating this to a similar question or concept you have answered. If this doesn't trigger anything then go to the next question. Mark a check by each question you have visited but not yet answered. When you return to those checked questions see if you have gained new insight into solving the question or can apply a guessing strategy. If questions are all worth the same, always allow a maximum time for each question on your first run through the test. (e.g. 60

questions in 60 minutes means one question per minute first run though). Do at least 4 time checks during the test.

ALWAYS TALK TO YOURSELF IN A POSITIVE WAY DURING THE TEST AND COACH YOURSELF ON THE QUESTION YOU ARE ANSWERING. YOU ARE YOUR BEST ACOLYTE!

WE NEED YOUR OPINION

Dear Student:

*Math Advancement's main goal is to prepare students like yourself for the best academic achievement possible. It is important for us to know how effective **Math Advancement** was in preparing you for a successful performance on your major exam.*

Therefore, we appreciate your completion of this questionnaire as soon as you receive your scores. Your answer to these questions will be completely confidential. However, if you wish to receive further information about products similar to Math Advancement, then fill in the information after the questionnaire.

Thank you for your participation.

Directions: Please answer the questions below **after** you receive your test scores.

1. What exam were you preparing for with **Math Advancement**? _________
2. Have you taken this exam prior to using **Math Advancement**? _________
3. If your answer to 2 was yes, what was your prior math score? _________
4. What score did you receive in Math after preparing with **Math Advancement**? _________
5. How would you describe the help that **Math Advancement** offered you? _________

   a) Very helpful b) Somewhat helpful c) Not helpful
6. Would you recommend **Math Advancement** to a friend who might be preparing for a similar exam? ___
7. How would you rank your mathematics confidence prior to using **Math Advancement**? _________

   a) Very little confidence b) some confidence c) much confidence
8. Please circle the mathematics classes that you have successfully completed:

   a) general mathematics b) algebra c) geometry d) trigonometry
   e) calculus f) statistics g) others ____________________

If you would like a brochure of other products by Dr. Marilyn Simon, please fill in the information below:

Name ______________________________

Address ______________________________

____

Telephone No. (   ) ______________________________

We appreciate your participation. Please return your completed questionnaire to
BEST-PREP 15050 Rancho Real, Del Mar, California 92014
or e-mail: msimon@waldenu.edu

FEEDBACK TO BEST-PREP INSTRUCTORS

Dear ______________________

Thank you for sharing your knowledge and skills with us. I'd like to share with you some of my reactions to your class:

1. These are a few words that describe my experience of your class:

2. These are some things I liked about you and your teaching style:

3. On my test I think I will find what I learned

   ❑ Not useful ❑ Kind of useful ❑ Very useful

   I am preparing to take the:
   ❑ GRE® TEST ❑ GMAT™ ❑ SAT™ ❑ PSAT™ ❑ NTE CORE BATTERY™
   ❑ NTE MATH™ ❑ ACT™ ❑ CBEST™

4. The part or parts of the course I think I will find most useful are:

5. The part or parts I liked **best** were:

6. If I could offer you one or more suggestions that would make your class more exciting and useful to me it would be:

7. Here's a personal note:

8. Would you recommend this class to a friend?

Location of Workshop__________Date__________

Pre-Test Score__________Post-Test Score__________

Confidence level (from 1 - 10) ______________

❑ Male ❑ Female

THANK YOU!

ANSWER KEY

Page

11 (Pre-Test) 1) d 2) e 3) b 4) d 5) e 6) c 7) d 8) b 9) e 10) d
14 1) c 2) c 3) c
17 1) (n – m)/mn 2) b 3) c 4) –4 5) d 6) d 7) a 8) a
20 1) 5/10 = 2/4, 10/5 = 4/2, 2/5 = 4/10, 5/2 = 10/4
22 1) x = 22/7 2) y = 36/7 3) ? = am/b 4) ? = QRX/10T 5) w = 23 6) x = 2
24 1) 270 2) b 3) b 4) mw/d 5) c 6) $1.05
26 1) c 2) –3/4 3) $3/\sqrt{5}$ 4) 7/15, 4/9, 3/7, 5/12 5) 2^{17}
28 1) b 2) 12.5 3) $20 4) e 5) b 6) b
30 1) $5\sqrt{3}$ 2) $3x\sqrt{y}$ 3) 75 4) .01 5) m^{50} 6) $13k^5\sqrt{k}$ 7) d 8) a
32 1) 2 2) –2 3) 0.21, 0.22, etc. 4) RS = –5 5) d 6) b 7) e 8) b 9) 3N + 7 = 16, N = 3
10) 2N – 4 = 10, N = 7
35 1) b 2) 35 3) a 4) b 5) 2^{17} 6) a 7) 16 8) d 9) a
40 1) 57 2) 50 3) 98 4) a 5) b 6) b 7) a 8) e 9) 41
43 (Check Point # 1): 1) d 2) d 3) d 4) a 5) a 6) b 7) a 8) b 9) a 10) c 11) b
12) a 13) c 14) a 15) c 16) b 17) b 18) 81 19) d 20) c 21) b 22) b
23) c 24) c 25) 72% 26) c 27) 105 28) 5
49 1) a 2) b 3) 72 4) b 5) a 6) c 7) d
52 1) 20 2) c 3) c 4) b 5) 9 6) 8.1 or any x, where: $8.0 < x < 9.0$
56 1) b 2) 100 π cm 3) b 4) a
58 1) x = 35, y = 25 2) d
60 Garfield, 1) 34 2) c
63 (Check Point # 2) 1) e 2) e 3) d 4) b 5) c 6) d 7) c 8) d 9) b 10) d 11) c 12) b
13) c 14) b
69 1) b 2) b 3) b 4) c 5) c
70 1) a 2) c 3) c 4) d 5) a 6) b
73 1) 40 2) 70 3) 110 4) 70 5) 70 6) 35 7) 85 8) 85 9) 70 10) 55 11) 110 12) 110
13) 50 14) 35 15) 60 16) 95 17) 55 18) 25 19) 35 20) 60 21) 55 22)120 23) 45
75 1) d 2) c 3) c 4) a 5) e 6) c 7) c 8) e
82 1) b 2) c 3) b 4) c 5) c 6) c 7) a 8) c 9) a
83 1) c 2) d 3) a 4) c
92 1) a 2) b 3) b 4) e 5) b 6) e
93 1) b 2) b 3) e 4) a 5) c 6) d
94 1) e 2) d 3) e 4) a 5) b 6) c
95 1) b 2) c 3) c 4) d 5) a 6) e 7) e
96 1) d 2) c 3) e 4) d 5) b 6) a
100 1) c 2) c 3) c 4) d 5) a 6) b 7) d
101 1) d 2) a 3) a 4) b 5) b 6) e
102 1) a 2) b 3) b 4) b 5) d
103 1) a 2) e 3) e 4) d 5) b 6) b
104 1) d 2) c 3) c 4) e 5) b 6) a 7) b
107 (Post-Test) 1) b 2) d 3) a 4) a 5) d 6) c 7) b 8) c 9) d 10) c